Springer Transactions in Civil and Environmental Engineering

Springer Transactions in Civil and Environmental Engineering (STICEE) publishes the latest developments in Civil and Environmental Engineering. The intent is to cover all the main branches of Civil and Environmental Engineering, both theoretical and applied, including, but not limited to: Structural Mechanics, Steel Structures, Concrete Structures, Reinforced Cement Concrete, Civil Engineering Materials, Soil Mechanics, Ground Improvement, Geotechnical Engineering, Foundation Engineering, Earthquake Engineering, Structural Health and Monitoring, Water Resources Engineering, Engineering Hydrology, Solid Waste Engineering, Environmental Engineering, Wastewater Management, Transportation Engineering, Sustainable Civil Infrastructure, Fluid Mechanics, Pavement Engineering, Soil Dynamics, Rock Mechanics, Timber Engineering, Hazardous Waste Disposal Instrumentation and Monitoring, Construction Management, Civil Engineering Construction, Surveying and GIS Strength of Materials (Mechanics of Materials), Environmental Geotechnics, Concrete Engineering, Timber Structures.

Within the scopes of the series are monographs, professional books, graduate and undergraduate textbooks, edited volumes and handbooks devoted to the above subject areas.

More information about this series at http://www.springer.com/series/13593

Nihal Anwar Siddiqui · S. M. Tauseef ·
Rajendra Dobhal
Editors

Advances in Water Pollution Monitoring and Control

Select Proceedings from HSFEA 2018

Springer

Editors
Nihal Anwar Siddiqui
Department of Health Safety,
Environment and Civil Engineering
University of Petroleum
and Energy Studies
Dehradun, Uttarakhand, India

S. M. Tauseef
Department of Health Safety,
Environment and Civil Engineering
University of Petroleum
and Energy Studies
Dehradun, Uttarakhand, India

Rajendra Dobhal
Uttarakhand State Council for Science
and Technology (UCOST)
Dehradun, Uttarakhand, India

ISSN 2363-7633 ISSN 2363-7641 (electronic)
Springer Transactions in Civil and Environmental Engineering
ISBN 978-981-32-9955-9 ISBN 978-981-32-9956-6 (eBook)
https://doi.org/10.1007/978-981-32-9956-6

This Springer imprint is published by the registered company Springer Nature Singapore Pte Ltd.
The registered company address is: 152 Beach Road, #21-01/04 Gateway East, Singapore 189721, Singapore

Committees

Steering Committee

Chief Patron

Dr. S. J. Chopra, Chancellor, UPES, Dehradun, India

Patron

Dr. Deependra Kumar Jha, Vice Chancellor, UPES, Dehradun, India

General Chairs

Dr. Kamal Bansal, Dean, School of Engineering (SoE), UPES, Dehradun, India
Prof. F. I. Khan, Professor and Canada Research Chair (Tier I)
Director, Centre for Risk, Integrity and Safety Engineering (C-RISE)

Programme Chair

Prof. S. A. Abbasi, Emeritus Professor (CSIR), Centre for Pollution Control and Environmental Engineering, Pondicherry University, Pondicherry, India

Organizing Chairs

Prof. Suresh Kumar, Director (SoE), UPES, Dehradun, India
Prof. Manish Parteek, Director (SoCS), UPES, Dehradun, India
Prof. Rajnish Garg, Professor, UPES, Dehradun, India

Organizing Co-chairs

Dr. N. A. Siddiqui, HOD (HSE), UPES
Dr. S. M. Tauseef, Associate Professor (HSE), UPES

Publicity Chairs

Dr. N. Siddiqui, HOD (HSE), UPES
Dr. B. P. Yadav, Associate Professor (HSE), UPES

Public Relation Chairs

Dr. Jitendra Kumar Pandey, Head (R&D), UPES
Prof. P. Mondal, A.P. (SG), HSE, UPES
Prof. Akshi Singh, A.P. (SS), HSE, UPES

Session Management Chairs

Prof. P. Mondal, A.P. (SG), HSE, UPES
Dr. Kanchan Bahukhandi, A.P. (SG), HSE, UPES
Ms. Madhuben Sharma, A.P. (SS), HSE, UPES
Mr. Abhishek Nandan, A.P. (SS), HSE, UPES
Mr. Arun P. A., A.P. (SS), HSE, UPES

Organizing Committee

Convener

Dr. N. A. Siddiqui, Professor and HOD (HSE and Civil), UPES

Co-convener

Dr. B. P. Yadav, Associate Professor, HSE and Civil Engineering Dept., UPES

Conference Secretary

Dr. S. M. Tauseef, Associate Professor and Assistant Dean (R&D), UPES

Finance Committee

Mr. Prasenjit Mondal, A.P. (SG), HSE and Civil, UPES
Mr. Dharani Kumar K., A.P., HSE and Civil, UPES

Technical Support

Dr. S. M. Tauseef, Associate Professor and Assistant Dean (R&D), UPES
Dr. Kanchan Deoli Bahukhandi, A.P. (SG), UPES
Dr. Rajesh Singh, Associate Professor and Head, Institute of Robotics Technology (R&D), UPES
Ms. Anita Gehlot, A.P., UPES
Dr. Bhawna Yadav Lamba A.P., UPES

Dr. Shiley Singhal, Professor, Department of Applied Sciences, UPES
Dr. Neeraj Anand, UPES
Dr. Rajnish Garg, UPES
Dr. D. K. Gupta, Professor, Department of Petroleum Engineering, UPES
Dr. Manish Prateek, UPES
Dr. Vikash Garg, UPES
Dr. Suresh Kumar, UPES
Dr. OM Parkash, Department of Aerospace Engineering, UPES
Dr. Pankaj, Sharma, Professor, Dept. of Mechanical Engineering, UPES
Dr. Rajeev Gupta, Associate Professor, Department of Physics, UPES
Dr. Tarun Dhingra, Professor, College of Management, UPES
Dr. Prasoom Dwidi, Professor, College of Management, UPES

Organizing Committee

Dr. Kamal Bansal, Dean, SoE, UPES, Dehradun, India
Dr. Manish Parteek, Dean (SoCS), UPES, Dehradun, India
Dr. Rajnish Garg, Professor, UPES
Dr. Tabrez Ahmad, Director, CoLS, UPES, Dehradun, India
Dr. Suresh Kumar, Director, SoE, UPES
Dr. Shyamal Kumar Banerjee, Associate Dean—Planning and Monitoring, SoE, UPES
Dr. Jitendra Kumar Pandey, Associate Dean, R&D, SoE, UPES
Dr. Neeraj Anand, Professor, UPES
Dr. Nihal Anwar Siddqui, HOD, HSE and Civil Engineering Department, UPES
Dr. S. M. Tauseef, Assistant Dean (R&D), UPES
Dr. Bikrama Prasad Yadav, UPES
Mr. Prasenjit Mondal, UPES
Dr. Kanchan Bahukhandi, UPES
Ms. Madhuben Sharma, UPES
Mr. Abhishek Nandan, UPES
Mr. Venkat Krishna Kanth, UPES
Mr. Mopidevi Vijai Kishore, UPES
Mr. Ashish Yadav, UPES
Mr. Durga Prasad Pandey, UPES
Mr. Rahul Silori, UPES
Mr. Susanta Kumar Sethy, UPES
Mr. Surender V., UPES
Mr. Arun P., UPES
Mr. Akshi Singh, UPES
Dr. Ashutosh Gautam, GM, India Glycols Limited
Dr. R. K. Sharma, GM HSE, India Glycols Limited
Er. R. K. Singh, Deputy Chief Inspector of Factories

Member of International Technical Panel

Technical Committee for the Conference:
Dr. Salim Ahmed, Memorial University, Canada
Dr. Syed Imtiaz, Memorial University, Canada
Dr. Hossam Gaber, UOIT, Canada
Dr. Rouzbeh Abbassi, Macquarie University, Australia
Dr. Vikram Garaniya, University of Tasmania, Australia
Dr. Arshad Ahmed, University Technology Malaysia, Malaysia
Dr. Ming Yang, Nazerbev University, Kazakhstan
Dr. Risza Rusli, University Technology PETRONAS, Malaysia
Dr. Hamed Saber, Jubail University College, Saudi Arabia
Dr. Azmi Mohd Shariff, University Technology PETRONAS, Malaysia
Prof. A. Gairola, Indian Institute of Technology (IIT) Roorkee, India
Prof. S. A. Abbasi, Pondicherry University, India
Prof. F. I. Khan, Memorial University, Newfound Land, Canada
Prof. A. S. Rangwala, Worcester Polytechnic Institute, USA
Prof. A. Tugnoli, Universitá Di Bologna, Italy
Prof. D. Rashtchian, Sharif University of Technology, Iran
Prof. D. S. Arya, Indian Institute of Technology (IIT) Roorkee, India
Prof. H. Gaber, University of Ontario, Canada
Prof. I. D. Mall, Indian Institute of Technology (IIT) Roorkee, India
Prof. I. M. Mishra, Indian School of Mines (ISM) Dhanbad, India
Prof. Javed Ahmad, Dean Faculty of Science, Jamia Hamdard, India
Prof. K. Kesava Rao, Indian Institute of Sciences (IISc) Bangalore, India
Prof. R. Abbassi, University of Tasmania, Australia
Prof. S. Ahmed, Memorial University, Newfound Land, Canada
Prof. S. Ayatollahi, Sharif University of Technology, Iran
Prof. S. Mukherji, Indian Institute of Technology (IIT) Bombay, India
Prof. S. N. Upadhyay, Former Director, Indian Institute of Technology (IIT) BHU, Varanasi, India
Prof. S. Shariatipour, Coventry University, UK
Prof. Shephard Ndlovu, University of Central Lancashire, UK
Prof. Sumeet S. Aphale, University of Aberdeen, UK
Prof. Suresh Jain, Teri University, New Delhi, India
Dr. Anurag Garg, Indian Institute of Technology (IIT) Bombay, India
Dr. Asha Rajvanshi, Professor and Head, Wildlife Institute of India, India
Dr. B. Abdolhamidzadeh, Sharif University of Technology, Iran
Dr. B. Sengupta, Former Member Secretary, Central Pollution Control Board (CPCB), India
Dr. C. Rezzaei, KOC, Kuwait
Dr. Elangawan, Deputy Director, Director General Factory and Services Labour Institute, Mumbai, India
Dr. M. K. Chandel, Indian Institute of Technology (IIT) Bombay, India

Preface

Depleting freshwater resources, shrinking glaciers and erratic rain patterns are all indicators of impending acute freshwater shortage that humanity is going to face if a course correction is not done immediately. Water is central to sustainable development and is critical for socio-economic development, healthy ecosystems and for sustaining life on earth itself.

As per *The United Nations World Water Development Report 2018: Nature-based Solution for Water*, nearly 3.6 billion people (approximately half the global population) are already living in potential water-scarce areas at least one month per year and this could increase to 4.8–5.7 billion in 2050. Similarly, the number of patients reporting health problems related to consumption or because of coming in contact with contaminated water is increasing every year.

Through this publication, the reader can update himself/herself with the advances in water pollution monitoring and control and inform on related opportunities and challenges. This volume presents select papers on advances in water pollution monitoring and control which were presented at the 2nd International conference on advances in the field of health, safety, fire, environment, allied sciences and engineering (HSFEA 2018), 16–17 November 2018 organized in collaboration with the Centre for Risk, Integrity and Safety Engineering (C-RISE), Memorial University, Canada. The conference was attended by leading academic scientists, leading engineers, policy makers, budding scholars and graduate students. The contribution from the authors cover topics ranging from green technology that would help tackle the problem of water pollution and ensure access to water safe for consumption and other uses to methods and means that need to be adopted to ensure sustainable development. Topics on the effect of climate change on water availability and usage are also presented.

Dehradun, India S. M. Tauseef

Acknowledgements

This volume has benefitted from contributions from participants at the 2nd International conference on advances in the field of health, safety, fire, environment, and allied sciences (HSFEA 2018), November 16–17, 2018 organized in collaboration with the Centre for Risk, Integrity and Safety Engineering (C-RISE), Memorial University, Canada, and the support and input of a number of individuals and institutions.

We thank Dr. S. J. Chopra (Hon'ble Chancellor, UPES) and Prof. Dr. Deependra Kumar Jha (Vice Chancellor, UPES), for their support and encouragement. We are grateful to the Chief Guests for HSFEA 2018—Mr. Howard Pike, (Former) Manager, Operations and Chief Safety Officer, Newfoundland Offshore Petroleum Board, Canada and Prof. Faisal Khan, Professor and Canada Research Chair (Tier I), Director, Centre for Risk, Integrity and Safety Engineering (C-RISE), Memorial University of Newfoundland, Canada—for gracing the event with their presence. Distinguished speakers—Senior Prof. S. A. Abbasi (CSIR Professor Emeritus, Pondicherry University), Mr. Satya Prakash Garg (Executive Director, GAIL (India) Limited), Mr. Rajendra Singh (Water man of India, Water Conservationist and Environmentalist), Dr. Tasneem Abbasi (Assistant Professor, Pondicherry University), Dr. Niranjan Bagchi (Former Director, MoEF), Dr. T. K. Joshi (Advisor, Environmental Health, Ministry of Environment, Forest and Climate Change) and Dr. R. K. Sharma (General Manger, India Glycols Ltd.) for their talks.

The organizers of HSFEA 2018 wish to thank all the reviewers for their valuable time and comments on the quality of the papers.

We acknowledge the support of our sponsors Council of Scientific and Industrial Research (CSIR), New Delhi; Atomic Energy Regulatory Board (AERB); GAIL; Uttarakhand State Council for Science and Technology (UCOST); Department of Labour, Uttarakhand State; Uttarakhand Jal Sansthan; Action For Sustainable, Efficacious Development and Awareness (ASEA); PARAM Environmental Solutions; Pratham Hospitality Services (PHS); Britannia; SWAN Scientific and Akbar HSE.
We also thank the chairs and members of various committees.

Contents

About the Editors

Nihal Anwar Siddiqui is Professor at the University of Petroleum and Energy Studies (UPES) (2005–present). He holds a Master's degree in Environmental Sciences from Lucknow University, India (1996), and a PhD in Environmental Biology from Awadhesh Pratap Singh University (APS) University, Rewa, India. Additionally, he has a number of professional certifications, e.g., a NEBOSH diploma, NEBOSH–IGC, IOSH Managing Safely, Lead Auditor OHSAS 18001, Lead Auditor ISO 14001 and GRAD-IOSH. Prof. Siddiqui's research interests include occupational health, safety and environment. He has co-authored more than 100 papers and eight books and has organized a number of national and international seminars/conferences/MDPs. He has also worked on a number of major R&D projects.

S. M. Tauseef is Associate Professor and Assistant Dean (R&D) at the University of Petroleum and Energy Studies (UPES), India (2015–present). He holds an MS in Chemical Engineering (Process Design) from Sharif University of Technology, Iran (2008), and a PhD in Environmental Engineering from Pondicherry University (PU) (2011). Prior to joining UPES, he worked as a Pool Officer (CSIR, Council of Scientific and Industrial Research) at PU (2012–2015). Dr. Tauseef's research interests include process safety, especially forecasting and consequence assessment of accidents in chemical process industries. He has developed methodologies for the application of computational fluid dynamics (CFD) to dense gas dispersion, vapor cloud explosions, and single and multiple pool fires. His other main interest is in solid waste management, especially the design of continuously operable vermireactor systems for assorted wastes, and high-rate anaerobic digesters. To date, he has published 60 journal articles, co-authored three books, published four patents and presented papers at 21 conferences.

Rajendra Dobhal is the Director General of the Uttarakhand State Council for Science and Technology, Govt. of Uttarakhand, India. His previous positions have included serving as the Chairman and Managing Director of the National Research Development Corporation (NRDC), Govt. of India, the Director of the Uttarakhand

Science Education and Research Centre (USERC) and the Uttarakhand Space Application Centre (USAC), among others. Dr Dobhal has monitored over 250 R&D projects leading to national/ international publications and patents and successfully organized 11 science congresses in Uttarakhand leading to benefit over 10,000 scientists. He has written over 40 technical reports, 12 books, published over 150 research papers in various journals of national & international repute.

Anchorage System for Penstocks—Case Study of Champawat Town Pumping Water Supply Scheme

A. N. Shankar, Imran Hussain and S. C. Gupta

Abstract This paper deals with the design of anchorage system for penstock of the proposed palatable water supply line for the reorganisation scheme of Uttarakhand Peyjal Nigam for Champawat town, Champawat district in state of Uttarakhand, India. The penstock has deviations and change in the elevation. Because of this, the centrifugal force and unbalanced internal pressure is generated by moving water which will produce the horizontal thrust on the penstock bends and at joints. Since the pressure in the penstock is varying with respect to change in direction and elevation which in turn lead to bursting or rupture of the penstock. Hence for the economic usage of penstock, an anchorage system has been provided by means of RCC thrust block. This anchor provided to penstock will act as the thrust restraint due to its dead weights placed opposite to penstock's line of horizontal thrust. The design of the RCC thrust block has been done manually.

Keywords RCC thrust block · Momentum · Thrust absorber · Deviation angle · Penstock and palatable water

Notations

P	Thrust
P_f	Frictional resistance
P_p	Passive pressure
P_a	Active pressure

A. N. Shankar (✉) · I. Hussain · S. C. Gupta
Department of Civil Engineering, University of Petroleum and Energy Studies, Bidholi via Prem Nagar, Dehradun 248007, India
e-mail: anshankar@ddn.upes.ac.in

I. Hussain
e-mail: imranhssn159@gmail.com

S. C. Gupta
e-mail: scgupta@ddn.upes.ac.in

N. A. Siddiqui et al. (eds.), *Advances in Water Pollution Monitoring and Control*, Springer Transactions in Civil and Environmental Engineering,
https://doi.org/10.1007/978-981-32-9956-6_1

P_r	Total resistance in kN
p	Internal pressure of water in kN/m^2
A	Cross sectional area of pipe in m^2
w	Unit weight of water in kN/m^3
v	Velocity of flow of water in m/s
g	Acceleration due to gravity 9.81 m/s^2
θ	Deviation angle of pipeline in degrees
HYSD	High yield strength deformation bars
W	Total weight carried by thrust block in kN
W_b	Weight of thrust block in kN
W_w	Weight of water in pipe in kN
WS	Weight of soil cover in kN
K_p	Co-efficient of passive earth pressure
K_a	Co-efficient of active earth pressure
h	Height of earth mass which is equal to dia of pipe
H	Height of thrust block in m
L	Length of soil thrust block in m
B	Breadth of soil thrust block in m
Ø	Angle of internal friction of soil. It is the max inclination of a plane at which body remains in equilibrium over inclined plane by the resistance of friction only
C	Unit cohesion
γ_s	Unit weight of soil cover in kN/m^3
CPHEEO	Central Public Health and Environmental Engineering Organisation
T.E.L	Total energy line
GL	Ground level
RSF	Rapid sand filter
BFG	Boulder filled gallery
IPS	Intermediate pumping station
WTP	Water treatment plant
MS.ERW	Mild steel Electric resistance welded pipe
API.5L	Pipe specification as per ISO: 3183

1 Introduction

Champawat is a town and a nagar palika parishad in Champawat district in the state of Uttarakhand, India. Champawat town elevation is 1615 m above mean sea level, having a total area of 5 km^2 and its coordinates are 29.33°N 80.10°E (Wikipedia: Champawat, Uttarakhand, India). This town has a population of 30,000 (I.S.5330 1984). This scheme name is Reorganisation scheme which was launched by the Uttarakhand Peyjal Nigam (I.S.5330 1984). Since the residents are reeling under an acute water crisis as rain has damaged the penstock line of Champawat town, hence a new water supply line being layed for this town. The previous penstock line had got damaged due to the improper knowledge of laying of penstock, and the anchorage system had not been provided. Source of water is from the local Kuerala Nadi (river) which will finally meet Kai Nadi (river), Nepal. Water is conveyed to the existing line of 9 km within the town. Water after collection passed through BFG and later purified in RSF unit then supplied to the town through pumping in five stages from IPS within the total length 17 km raising main. The total cost of project is nearly 30 crore ₹. Water load coming on wtp is 4000 lpm and treated water going out for town is 3600 lpm (I.S.5330 1984). Hence design of anchorage blocks for the five stages of ips has been done in this paper (Figs. 1, 2 and 3).

Fig. 1 India political map (Neher and Lin 1978)

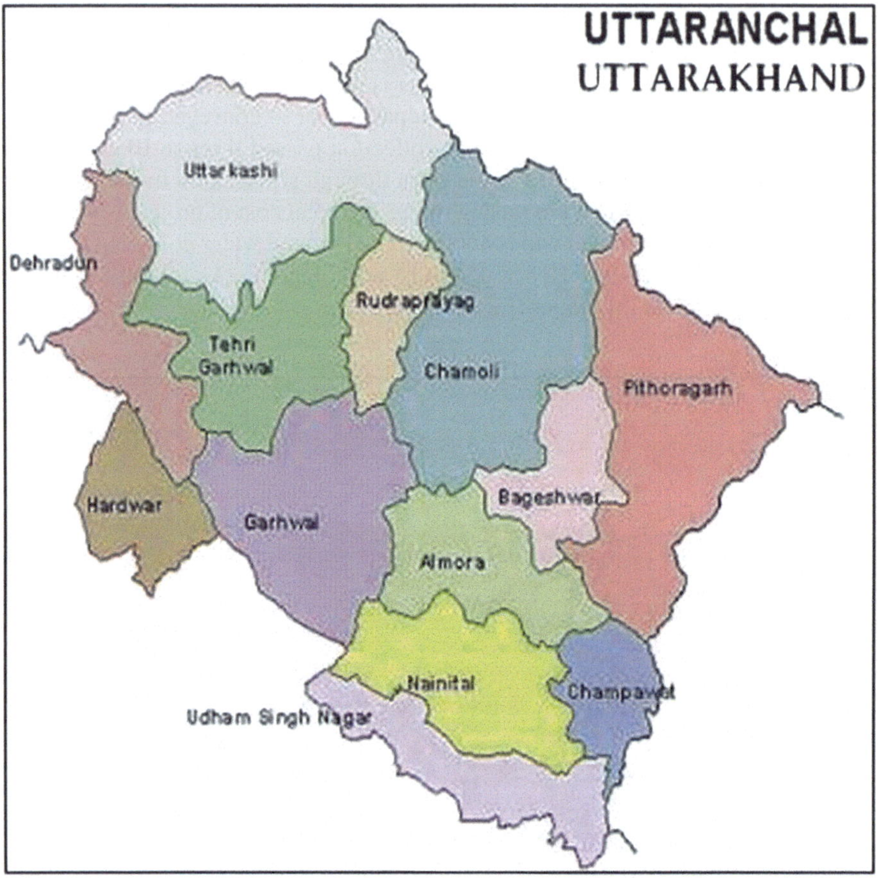

Fig. 2 Uttarakhand political map (Neher and Lin 1978)

2 Materials Requirement

2.1 Concrete

Concrete is a unique diversity material which is obtained by the mixture of inert materials which are called as aggregates, the cement and water. By controlling the proportion of these three components with which desired grade of M20 has been prepared and the minimum grade of the concrete should be M20 and maximum quantity of cement should be 560 kg/m^3 which has to be maintained (Political maps of India, Uttarakhand and Champawat are taken from Google). The water using for the manufacturing of the concrete was tap water (Political maps of India, Uttarakhand and Champawat are taken from Google; IS 456 2000; Punmia et al. 2014). The grade of concrete, quantity of cement and quality of water has been maintained at site.

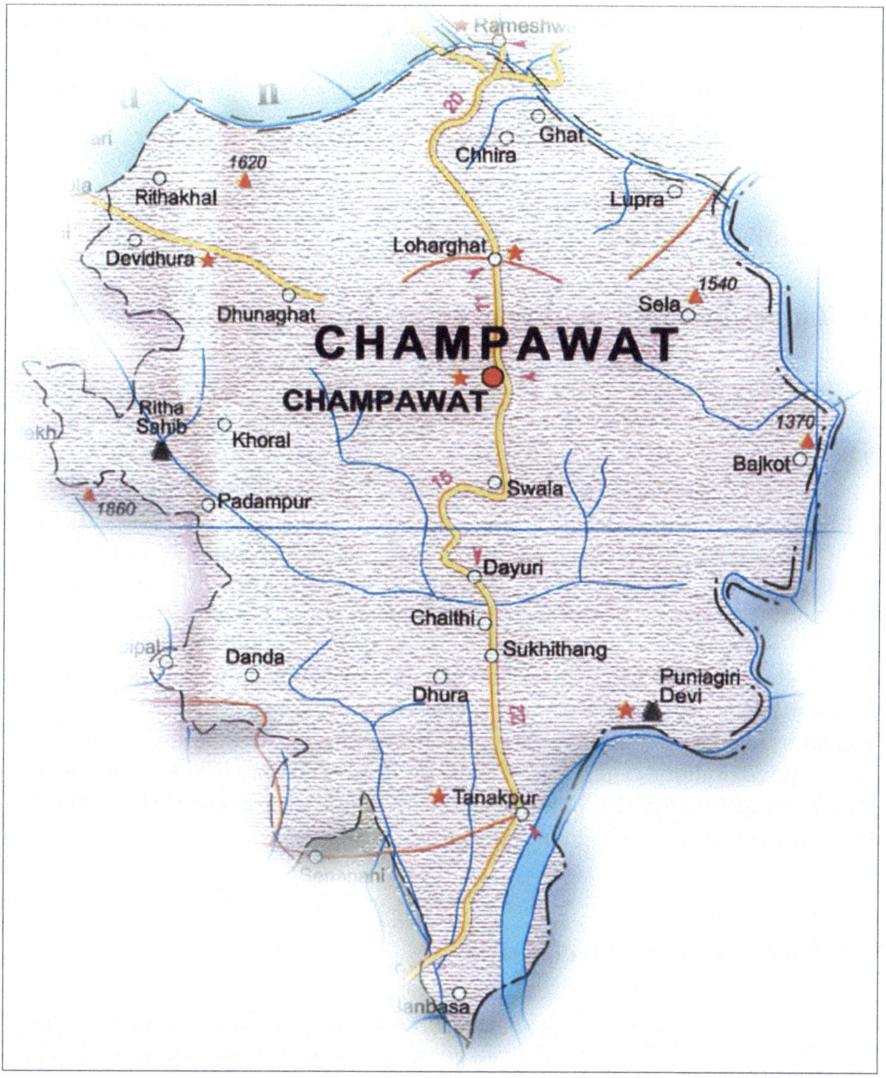

Fig. 3 Champawat political map (Neher and Lin 1978)

2.2 *Steel Reinforcement*

The concrete is strong in compression and weak in tension (Political maps of India, Uttarakhand and Champawat are taken from Google; IS 456 2000; Punmia et al. 2014), hence concrete was reinforced with the HYSD bars to take up the tension and to prevent concrete from the shrinkage and creep which will occur due to the variation of temperature with the time (Political maps of India, Uttarakhand and

Champawat are taken from Google; IS 456 2000; Punmia et al. 2014). Minimum area of steel used was 0.12% of surface area of anchorage block (Political maps of India, Uttarakhand and Champawat are taken from Google; IS 456 2000; Punmia et al. 2014).

3 Anchorage System Importance

3.1 Importance of Anchorage

The CPHEEO states "Anchorages are necessary to resist tendency of the pipes to pull apart at bends or other points of unbalanced pressure, or when they are laid on steep gradients and the resistance of their joints to longitudinal (shearing) stresses is either exceeded or inadequate. They also used to restrain the expansion and contraction of rigidly joined pipes under the influence of temperature changes" (Vishwanath 2012).

There is a generation of thrust at the penstock bends and at the joints due to the flow of water. Due to this there is generation of overturning and unbalanced stress in the penstock (Birdie and Birdie 2015; Punmia et al. 2016; Garg 2010; Neher and Lin 1978; Krishna and Jain 2015; Manganaro 1968; Schmid and Bauhahn 1960; Vishwanath 2012; CPHEEO 1999; Uttarakhand Peyjal Nigam, India). To overcome this problem anchorage plays its vital role. Hence the resistance to the thrust is obtained by the friction of pipe exterior and by bearing value of soil in which the block is buried (Punmia et al. 2016; Garg 2010; Neher and Lin 1978; Krishna and Jain 2015; Manganaro 1968; Schmid and Bauhahn 1960; Vishwanath 2012; CPHEEO 1999; Uttarakhand Peyjal Nigam, India).

3.2 Force Transfer Mechanism

Unbalanced forces in a pipeline are developed at elbows, tees, wyes, reducers, valves, and dead ends (Punmia et al. 2016; Garg 2010; Neher and Lin 1978; Krishna and Jain 2015; Manganaro 1968; Schmid and Bauhahn 1960; Vishwanath 2012; CPHEEO 1999; Uttarakhand Peyjal Nigam, India). These forces are shown in Fig. 4

The force generated by moving water which is occurred at the penstock joint and bend is transferred to the thrust block and thrust block or thrust absorber transfer the thrust to surrounding soil in which it is buried. The load transferred to soil should be within the SBC of the soil (Punmia et al. 2016; Neher and Lin 1978; Krishna and Jain 2015; Manganaro 1968; Vishwanath 2012; CPHEEO 1999; Uttarakhand Peyjal Nigam, India).

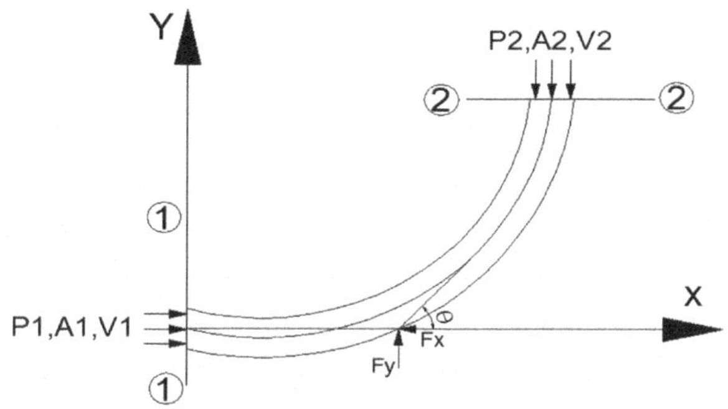

Fig. 4 Showing occurrence of thrust

4 Design of R.C.C Thrust Block

4.1 Design Data for Champawat Thrust Block

1. Internal pressure of penstock at 0 m chainage, p: 5687 kN/m^2
2. Angle of internal friction, Ø: 30°
3. Velocity of water in pipe at 0 m chainage, V: 4 m/s
4. Density of soil (γ_s): 18 kN/m^3
5. Density of concrete (γ_c): 25 kN/m^3
6. Density of water (γ_w): 10 kN/m^3
7. Co-efficient of friction between soil and concrete, μ: 0.4
8. Cohesion of sandy soil, c: 0
9. Dia of penstock mentioned in below profile levelling data for different chainage
10. Soil cover: 1.0 m.

4.2 Profile Levels

See Figs. 5, 6, 7, 8 and 9.

4.3 Calculation of Thrust (P)

There is momentum as the water is flowing in penstock, when there is change in direction of flow momentum gets varies, Newton 2nd law states "the rate of change of momentum is directly proportional to the force applied in direction of force" (Garg 2010; QAWSM 1993). The resolution of the forces has been widely studied by reference (QAWSM 1993). The change in direction of penstock is measured as

Fig. 5 Showing stage-I levels

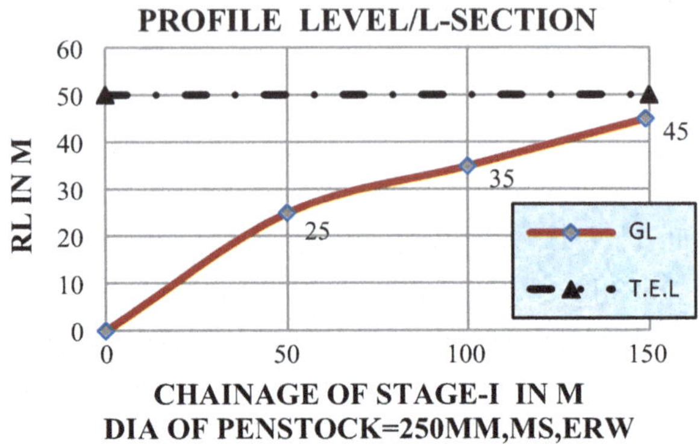

Fig. 6 Showing stage-II levels

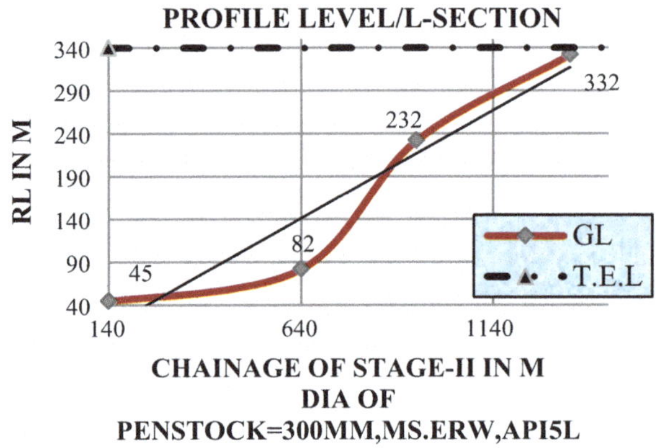

Fig. 7 Showing stage-III levels

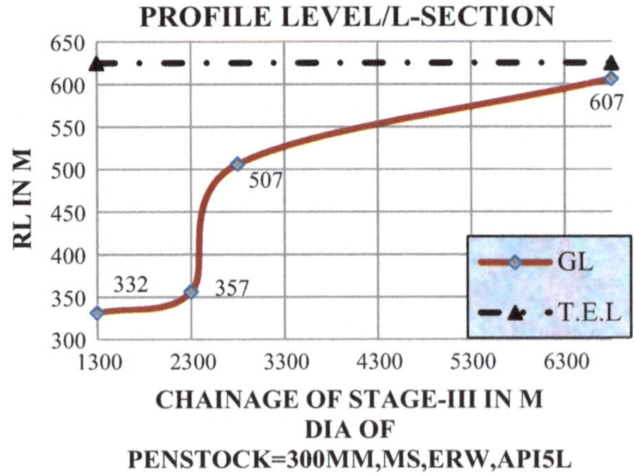

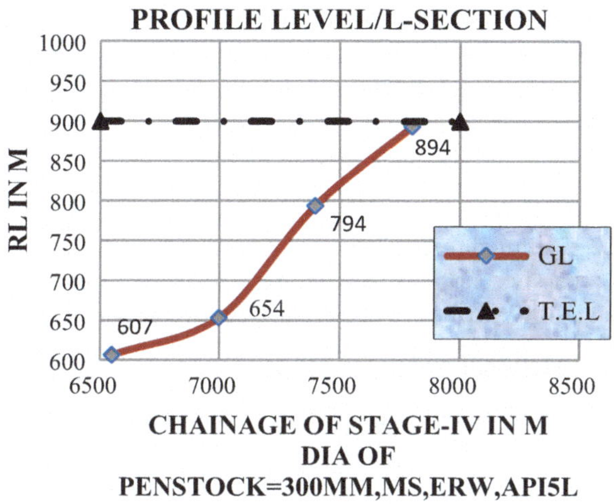

Fig. 8 Showing stage-IV levels

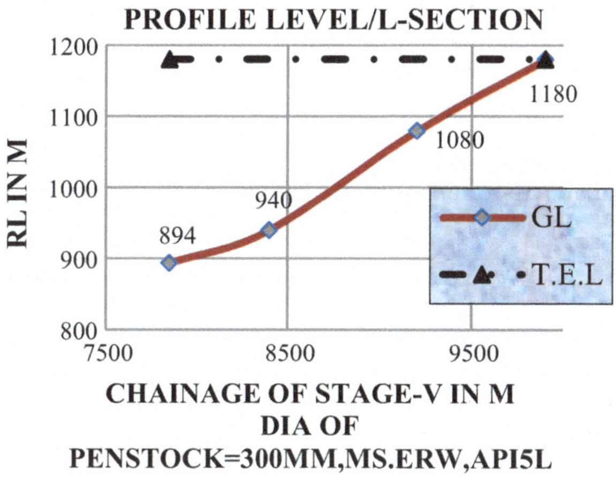

Fig. 9 Showing stage-V levels

the deviation angle in terms of degrees. Here the deviation angles are obtained by help of above surveying data of levels.

Applying momentum equation and resolving it, thrust is obtained as below

$$P = 2A(\frac{\gamma_w V^2}{g} + p)\sin\frac{\theta}{2}. \text{ And unit is in kN.}$$

4.4 Calculation of Resistances

The resistance to thrust offered are by the friction of penstock exterior and by bearing value of soil in which block is buried.

The total resistance offered by block contain 3 components (Punmia et al. 2016; Vishwanath 2012; CPHEEO 1999).

1. Frictional resistance.
2. Passive earth pressure (Lateral resistance of soil).
3. Active earth pressure (Lateral earth pressure).

a. **Frictional resistance (P_F):**

Resistance provided by the thrust block. The unit is in kN.

$$P_f = \mu W.$$

$$\text{and } W = W_b + W_w + W_S$$

b. **Passive pressure (P_p):**

It is the lateral resistance against the outer face of the pipe and block (CPHEEO 1999). The maximum pressure exerted by soil opposite to the thrust on the block is called passive pressure; the unit is in kN (Punmia et al. 2016; CPHEEO 1999).

$$P_p = \left(\gamma_s\left(\frac{H^2}{2}\right)L\left(\frac{1+\sin\emptyset}{1-\sin\emptyset}\right)\right) + \left(2CHL\sqrt{\frac{1+\sin\emptyset}{1-\sin\emptyset}}\right)$$

Or

$$P_p = \left(\gamma_s\left(\frac{H^2}{2}\right)LK_p\right) + \left(2CHL\sqrt{K_p}\right)$$

c. **Active pressure (P_a):**

It is the minimum pressure which may occur on thrust block is called active pressure (Punmia et al. 2016; CPHEEO 1999). Active earth pressure is developed if the thrust block is free to yield away from soil mass (Punmia et al. 2016; CPHEEO 1999). Will act on the portion of projected pipe i.e., when block is free to yield or slide away from soil. The unit is in KN.

$$P_a = \left(\gamma_s h\left(\frac{1-\sin\emptyset}{1+\sin\emptyset}\right)\right) - \left(2C\sqrt{\frac{1-\sin\emptyset}{1+\sin\emptyset}}\right)$$

Or

$$P_a = (\gamma_s h K_a) - \left(2C\sqrt{K_a}\right)$$

d. **Total resistance (P_r):**

It is the total resistance offered to the thrust (Punmia et al. 2016; CPHEEO 1999). It is the total summation of the frictional resistance, passive pressure and active pressure. The unit is in kN.

Therefore, Total resistance, $P_r = P_f + P_p + P_a$

e. **Change in pressure**

Due to change in elevation there is change in internal pressure in penstock. The change in pressure was calculated by applying the Bernoulli's continuity equation which has been studied from (Gupta 2017),

$$\left(\frac{p_1}{\gamma_w g}\right) + \left(\frac{V_1^2}{2g}\right) + Z_1 = \left(\frac{p_2}{\gamma_w g}\right) + \left(\frac{V_2^2}{2g}\right) + Z_2$$

f. **Factor of safety (FOS):**

Factor of safety is the ratio of ultimate strength to the actual working stress or max permissible stress when in use (Punmia et al. 2014, 2016; IS 456 2000; Vishwanath 2012; CPHEEO 1999).

Here it is the ratio of the total resistance to the horizontal thrust. It should be greater than 2 and less than 3. If factor of safety is with in 2 and 3 it will safely carry the horizontal thrust and also the water hammer effect. If the factor of safety is within above prescribed limits then the thrust block will be efficient and economical for intended purpose of work (Punmia et al. 2016; CPHEEO 1999; Uttarakhand Peyjal Nigam, India).

$$FOS = \frac{P_r}{P}$$

FOS should be >2 and <3 for the optimum size of the thrust block (Punmia et al. 2016; CPHEEO 1999; Uttarakhand Peyjal Nigam, India).

5 Results and Discussion

The Table 1 shows the results which were obtained for sizes of thrust block for five stages of intermediate pumping stations. The obtained sizes were employed for the Champawat town pumping water supply scheme. The above obtained all thrust blocks has the FOS >2 and <3. Hence the obtained sizes of thrust blocks are safe for the working of penstock. And the surface steel reinforcement of 8Ø@200 mm c/c provided. Figure 10 shows the typical plan and section of anchorage block that is thrust block which is showing the reinforcement arrangement in the thrust block.

Table 1 Factor safety for various stages

Stage	Chainage in 'm'	Dia of penstock in 'm'	Thrust (P) in kN	Size of thrust block L × B × H in 'm'	FOS
I	0	0.25	126.178	2.0 × 2.0 × 2.0	2.49
	50	0.25	54.1064	1.5 × 1.5 × 1.5	2.50
	100	0.25	126.178	1.5 × 1.5 × 1.5	2.50
II	140	0.3	103.201	1.8 × 1.8 × 1.8	2.25
	640	0.3	204.636	2.2 × 2.2 × 2.2	2.00
	940	0.3	204.636	2.2 × 2.2 × 2.2	2.00
III	1300	0.3	103.201	1.8 × 1.8 × 1.8	2.25
	2200	0.3	103.201	1.8 × 1.8 × 1.8	2.25
	2900	0.3	103.201	1.8 × 1.8 × 1.18	2.25
IV	6600	0.3	103.201	1.8 × 1.8 × 1.18	2.25
	7000	0.3	204.636	2.2 × 2.2 × 2.2	2.00
	7400	0.3	103.201	1.8 × 1.8 × 1.8	2.25
V	7800	0.3	103.201	1.8 × 1.8 × 1.8	2.25
	8400	0.3	103.201	1.8 × 1.8 × 1.8	2.25
	9200	0.3	103.201	1.8 × 1.8 × 1.8	2.25

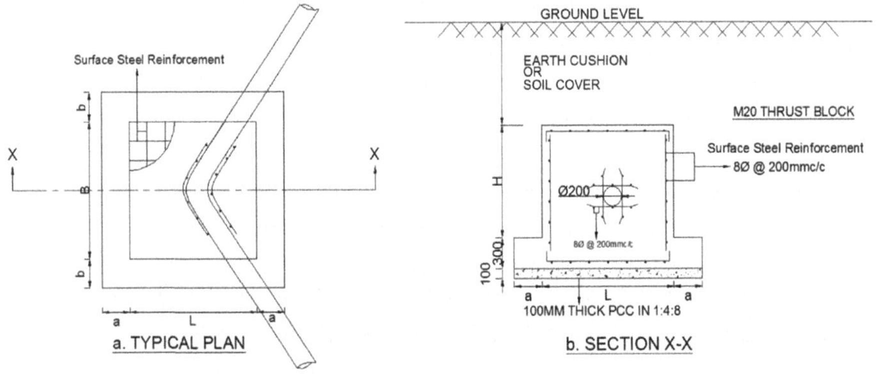

Fig. 10 Showing typical plan and section of thrust block

References

Birdie GS, Birdie JS (2015) Water supply & sanitary engineering, 9th edn. Dhanpat Rai Publishing Company, pp 110–112

CPHEEO (1999) Manual on water supply & treatment, 3rd edn. The Government of India, Ministry of Urban Development, New Delhi, pp 159–165, 609–611

Garg SK (2010) Environmental engineering, vol 1, Water Supply engineering, 20th edn. Khanna Publishers, pp 307–309

Gupta SC (2017) Engineering mechanics-1 (Statics &Dynamics). Nirali Prakashan, pp S 1.1–S 3.45
Handbook on pipes and fittings for drinking water supply-SP 57 (QAWSM) (1993)
IS 456 (2000) Plain and reinforced concrete—Code of practice
I.S.5330 (1984) Criteria for design of anchor blocks for penstocks with expansion joints [WRD 14: Water Conductor Systems]
Krishna J, Jain OP (2015) Plain and reinforced concrete, vol 1, 8th edn. Nem Chand & Bros, pp 1–38,146–159
Manganaro CA (1968) Design for unbalanced thrust for buried water conduits. Journal (American Water Works Association) 60(6):705–716
Neher GE, Lin YT (1978) Practical applications for design of thrust blocks and tied joints. Journal (American Water Works Association) 70(3):131–138. Groundwater: Influencing Factors
Political maps of India, Uttarakhand and Champawat are taken from Google
Punmia BC, Jain AK, Jain AK (2014) R.C.C designs, 10th edn. Laxmi Publications, pp 1–29
Punmia BC, Jain AK, Jain AK (2016) Environmental engineering-1, Water Supply engineering, 2nd edn. Laxmi Publications, pp 485–502
Schmid GG, Bauhahn EE (1960) Importance of thrust blocks for pipe installations. Journal (American Water Works Association) 52(1):81–87
Uttarakhand Peyjal Nigam, India. Complete Data from department
Vishwanath HS (2012) Hydraulics, 2nd edn. Sapna Publications, pp 50–78
Wikipedia: Champawat, Uttarakhand, India

An Overview on Toxicity, Detection, and Removal of Chromium in Water and WasteWater

Bhavtosh Sharma and Prashant Singh

Abstract The inorganic pollutants especially toxic metals in excess quantity deteriorate the quality of water. In current climatic conditions, the quality of water has become a serious issue to our society due to increase in anthropogenic as well as natural activities. Moreover, the amount of chromium in water resources has increased not only in most of the places in India but also abroad. Consequently, the toxic nature of chromium including its occurrence, geochemistry, detection, and removal techniques from water and wastewater have been studied in the present paper. The paper also describes the chemical nature of various forms of chromium.

Keywords Chromium · Toxic nature · Geochemistry · Detection · Water · Wastewater · Removal methods · Treatment

1 Introduction

Usually, heavy metals occur naturally with high atomic weight and a density of at least 5 times more than that of water. The toxic nature of heavy metals mainly depends on its amount, exposure route, and specific chemical form. The metal ion speciation study is the recognition of an individual form of an element along with its oxidation states in a sample (Gupta et al. 2012). The chemical speciation term was described by Ure (1990). The toxic nature, bioavailability, and bioaccumulation of a specific metal depend on its oxidation states (Ebdon et al. 2001) and the elemental speciation tool to identify the actual chemical species of an element along with its quantity have also been studied (Montes-Bayon et al. 2003; B'Hymer et al. 2000). The aquifers are being contaminated with these pollutants in larger quantities. As a result, the metal ion speciation study in drinking water and wastewater samples is very important to know the exact situation of pollution. The industrial, agricultural and, domestic practices

B. Sharma (✉)
Uttarakhand Science Education Research Center (USERC), Dehradun, Uttarakhand, India
e-mail: bhavtoshchem@gmail.com

P. Singh
Chemistry Department, DAV Post Graduate College, Dehradun, Uttarakhand, India

N. A. Siddiqui et al. (eds.), *Advances in Water Pollution Monitoring and Control*,
Springer Transactions in Civil and Environmental Engineering,
https://doi.org/10.1007/978-981-32-9956-6_2

release the circulation of toxic and heavy metals in the environment and impacts on living beings. The metal ion speciation analysis has been established as a significant analytical tool. Nowadays, it has been possible to detect trace amount of metals and their species up to lower level detection limits with advanced instrumentation. The present paper highlights the occurrence, geochemistry, and detection with removal techniques of chromium from water and wastewater.

2 Geochemistry and Toxicity of Chromium in Water

The occurrence of chromium in the groundwater system was supposed to be connected largely with anthropogenic contamination (Hem 1959). It possesses a broad range of industrial applications (Nriagu and Niebor 1988). However, Robertson reported that chromium occurs as hexavalent chromium, Cr(VI) in alkaline and oxic groundwater (Robertson 1975, 1991). It has been reported that Cr(VI) form is very much soluble and mobile in alkaline water and oxic groundwater as well as its contamination may go for several meters with only limited reduction (Perlmutter et al. 1963; Chowdhury et al. 2016). The common oxidation states of Cr are (0), (III), and (VI). Generally, chromium occurs in two oxidation states in the environment, i.e., trivalent chromium Cr(III) and hexavalent chromium Cr(VI). Mostly, Cr occurs in nature in Cr(III) form which is the most stable form. Cr(VI) is considered as harmful species even in small quantity. Cr(III) is an essential nutrient for humans in moderate ingestion and its shortage may cause heart problems, disruptions of metabolisms and diabetes. Besides this, Cr(III) is essential for the synthesis of fat from glucose and also for the oxidation of fat to carbon dioxide. But the uptake of too much Cr(III) can cause health effects like skin rashes. Cr(III) is strongly sorbed on to mineral surfaces while sparingly soluble below pH 7 as reported by Kotas and Stasicka (2000).

The oxidation of Cr(III) form to Cr(VI) form takes place in the presence of oxides of Manganese (Schroeder and Lee 1975; Eary and Rai 1987; Guha et al. 2000). Microbially mediated reduction of Cr(VI) to Cr(III) are found in the presence of organic matter as well as other reducing agents as described by Rai and Zachara (1984). Sometimes, the reduction of Cr(VI) to Cr(III) may occur abiotically (Beller et al. 2014). Cr(VI) is very dangerous to human health especially for the people working in the steel and textile industries.

The trivalent form of chromium is the most common naturally occurring state in soils and rocks having fewer quantities of chromic oxide (Cr_2O_3), whereas, hexavalent chromium occurs as chromates ($CrO_4{}^{2-}$) and dichromate ($Cr_2O_7^{-2}$) ions in the environment with industrial and domestic effluents. It has been established that Cr(VI) in leather products may cause allergic reactions like skin rashes, upset stomachs, and ulcers, respiratory problems, weakened immune systems, kidney and liver damage, alteration of genetic material, lung cancer, etc. (Water Research Foundation 2015; Central Water Commission 2014; IARC 2012). According to the International Agency for Research on Cancer (IARC 2012), the people living in the vicinity of anthropogenic sources of Cr(VI) may be exposed through inhalation of contaminated ambient air or ingestion of contaminated drinking water. Cr(VI) is carcinogenic if inhaled by anyone (Daugherty 1992; ATSDR 2012; Sedman et al. 2006a, b; Beaumont et al. 2008). The wide industrial uses may produce regional pollution like the contamination observed in alluvial aquifers in the Los Angeles metropolitan area as reported by USEPA (2008). Chromium is usually substituted within minerals like amphiboles, pyroxenes, garnets, spinels, and micas (Motzer 2005; Reimann and de Caritat 1998). Chromite ($FeCr_2O_4$) is the most common chromium-containing mineral and is very much resistant to weathering. Scientists (Mayer and Schick 1981; Charlet and Manceau 1992) have described that the reduced chromium can be removed from the solution after balancing the pH of the solution.

3 Chromium Detection Methods

Cr(VI) is the second most stable oxidation state of chromium. Usually, Cr(VI) hardly occurs in nature. Cr(VI) may occur as an impurity in Portland cement. Hexavalent chromium compounds are being used widely as pigments for textile dyes, paints, inks, plastics; leather tanning, wood preservatives; corrosion inhibitors; metal finishing and chrome plating, etc. Cr(VI) can be generated and released during casting, welding, and cutting operations like stainless steel operations (IARC 2012). Cr(VI) can be reduced to the more stable Cr(III) by using the oxidizable organic matter or reducing agents such as iron (IARC 2012). The analytical instruments like Inductively Coupled Plasma Mass spectrometry (ICP-MS) Absorption Spectrometry (AAS), Atomic Emission Spectrometry (AES), Atomic Fluorescence Spectrometry (AFS) are useful methods. The potential carcinogenicity of Cr(VI) as a contaminant in drinking water has been studied by some researchers. The concentration of chromium in contaminated waters due to anthropogenic activities and leaching of wastewater (e.g., from landfills sites) can lead to drinking water contamination (IARC 2012). As per the report of WHO (2011), the total chromium has been reported in the range of 0.05–0.2 μg/L by the application of atomic absorption spectroscopy (AAS). WHO reported that the total chromium concentrations in drinking water are generally less than 2 μg/L, but levels as high as 120 μg/L have also been studied (WHO 2011). Bahadir et al. (2015) found trace amounts of hexavalent chromium in drinking waters by dispersive microsolid-phase extraction using modified multi-walled carbon nanotubes combined with total reflection X-ray fluorescence spectrometry technique. They found that the minimum Cr(VI) content which can be detected in an aqueous solution was 3 $\mu g\ L^{-1}$ which is 20 times lower than the maximum hexavalent chromium content permissible in drinking waters as per the World Health Organization (WHO) standard.

In California, 38% of municipal drinking water sources showed Cr(VI) concentrations more than 1 μg/L (Sedman and coworkers 2006a, b). Yang and coworkers (2001) determined the Cr(VI) in drinking water by electro-kinetic flow analysis system and graphite furnace atomic absorption spectrometry. A research group of Catalani et al. (2015) reported Cr(VI) concentration in tap water samples during the study of Cr(III) and Cr(VI) in 19 tap waters from various parts of Italy and in 14 commercial mineral waters using a metal-free HPLC–ICP–MS instrumentation with a reversed-phase C8 column. This team found that Cr(VI) ranged from 10.88 to 0.1 μg/L while Cr(III) ranged from 0.33 to 0.1 μg/L in tap water samples whereas, in commercial mineral waters, Cr(III) and Cr(VI) were in the range of 0.11–0.80 μg/L and 0.25–3.4 μg/L, respectively. Kaprara and his team (2015) have found Cr(VI) in 600 tap water of small–medium sized communities of Greece. The researchers have observed higher Cr(VI) values in aquifers in alluvial and neogene sediments of serpentine and amphibolites. However, a nickel–aluminum layered double hydroxide (Ni–Al LDH) is used as a nano-sorbent for the speciation analysis of chromium by flame atomic absorption spectrometry (FAAS). They used this method for the speciation of Cr in different water samples (Abdolmohammad-Zadeh and Sadeghi 2012).

4 Chromium Removal Methods

According to the American Water Works Association (AWWA 2013), the Conventional treatment, i.e., coagulation or lime softening followed by filtration are not very capable for hexavalent chromium as it is for total chromium removal. Furthermore, some other methods like strong base anion-exchange (SBA), reduction coagulation filtration (RCF), and single-use weak base anion-exchange (WBA) methods are slightly more effective tools. Selvi et al. (2001) reported the removal of Cr(VI) from aqueous solution by adsorption onto activated carbon prepared from coconut tree sawdust. Campos et al. (1995) have reported that Bacillus sp. have a very high potential for Cr(VI) removal from wastewater. Gupta and Ali (2004) obtained 96–98% removal of Cr(VI) using sugarcane bagasse ash in batch and column operations. Adsorption of Cr(VI) from aqueous solution by various microorganisms are given in Table 1.

Harijan and Chandra (2016) applied the functionalization of graphene sheets with polyaniline (PANI) for the treatment of toxic hexavalent chromium and reported that the adsorption on adsorbent was governed by electrostatic interaction between anions of Cr(VI) and amino groups in this process. They found that the maximum adsorption capacity of PANI functionalized graphene for Cr(VI) removal was 192 mg/g and the

Table 1 Adsorption of Cr(VI) from aqueous solution by various microorganisms

Microorganism	References
Arthrobacter viscous (Fungi)	Silva et al. (2009)
Aspergillus niger (Fungi)	Khambhaty et al. (2009)
Staphylococcus xylosus (Bacteria)	Ziagova et al. (2007)
Pseudomonas sp. (Bacteria)	Ziagova et al. (2007)
Chlorella vulgaris (Algae)	Nourbaksh et al. (1994)
Pachymeniopis sp. (Algae)	Lee et al. (2000)

Cr(VI) elimination capacity decreases with increase in pH and decrease in the initial concentration of Cr(VI).

Kumari and coworkers (2015) synthesized magnetite nanospheres with hollow interiors using a simple, one-pot, and template-free solvothermal method with ferric chloride as the iron precursor for the removal of Cr(VI) from water. This research group has also studied the composition, surface properties, and morphology with the help of X-ray powder diffraction (XRD), energy dispersive X-ray fluorescence (EDXRF), Fourier-transform infrared spectroscopy (FTIR), surface area analysis, point of zero charge (pH_{pzc}), scanning electron microscopy (SEM), transmission electron microscopy (TEM), and magnetic moment determination. This team has reported that these mesoporous nanospheres have a $S_{BET} = 11.3\ m^2/g$ and a high saturation magnetization of 77.5 emu/g. Finally, this group has found that these magnetite nanospheres effectively remediated Cr^{6+} from water.

Pérez et al. (2015) found 95% Cr(VI) removal by using a nano-hydrotalcite supported on SiO_2 and Nano-HT/SiO_2 was experienced in Cr(VI) retention through anion-exchange mechanism. In this study, the adsorption of Cr(VI) on nano-HT/SiO_2 fits well to the Freundlich isotherm. Hackbarth and coworkers (2016) reported a method for the treatment of galvanic wastewaters containing Cr(VI) and Cr(III) using Macroalga *Pelvetia canaliculata* as a natural electron donor and cation exchanger. They found in their study that Cr(VI) was reduced to Cr(III) by contact with electron-donor groups of selected biomass. The Cr(VI) reduction was evaluated as a function of the biomass and Cr(VI) concentration, pH, and temperature. The treatment of the wastewater having Cr(VI) was examined in a column which was packed with raw *P. canaliculata*, and the highest Cr(VI) reduction of about 2.1 mmol/g was achieved by these scholars.

Agrafioti et al. (2014) used the biochars derived from three different substances like rice husk, the organic fraction of municipal solid wastes, sewage sludge as well as a sandy loam soil for Cr removal and found that the biochar from municipal solid wastes was the most efficient in removing Cr(VI) from water. The research group found that the materials examined removed more than 95% of the initial Cr(III). The removal of Cr(VI) was considerably lower. Sadyrbaeva (2016) studied the removal of Cr(VI) from aqueous solutions using a novel hybrid liquid membrane-electro-dialysis process using the solutions of tri-*n*-octylamine in 1,2-dichloroethane as the liquid membranes and found the total removal of Cr(VI) from 0.01 mol L^{-1} $K_2Cr_2O_7$ solution. Bhaumik and his research team (2013) used the fixed bed column of polypyrrole/Fe_3O_4 nanocomposite for Cr(VI) removal from water. After this experiment, the experimental results confirmed that the breakthrough curves were dependent on bed mass, initial Cr(VI) concentration, and flow rate. Korngold and his research group (2003) used strong-base anion-exchange resins for selective removal of Cr(VI) from tap water. The impacts of anion-exchange resin, concentrations of Cl^- and SO_4^{2-} in the feed water, the flow rate, and pH of the feed water were investigated. Stergioudi et al. (2015) prepared open-cell copper foams using a space holder technique and tested as filter-beds for the uptake and reduction of Cr(VI) in drinking water. They reported that column tests proved the absolute Cr(VI) removal with minimum Cu leaching. Atieh (2011) investigated the application of the activated carbon and the carbon nanotubes attached to the activated carbon as potential adsorbents to Cr(VI) from polluted wastewater. Under this study, some parameters like concentration of Cr(VI), adsorbent dose, pH of the solution, contact time, and agitation speed were optimized. He found the activated carbon, which was coated with carbon nanotubes, as an excellent adsorbent to Cr(VI) ions with the adsorption capacity of 9.0 mg/g.

Yue et al. (2009) applied the low-cost chemically activated carbon materials like pellet-600 (a pelletized carbon material with high mesoporous volumes and surface area) and PVA-300 (a composition of a high surface area carbon coating on a fiberglass mat substrate) for the removal of Cr(VI) from water and found that these adsorbent materials exhibit more effective removal efficiency of Cr(VI) from water in comparison to commercially available activated carbons. Besides this, it has also been found that pellet-600 showed much faster adsorption kinetics and a much higher adsorption capability for Cr(VI) from water samples.

5 Conclusion

The toxic nature as well as the geochemistry of the chromium has been presented which is clearly indicating the toxic and nontoxic species of chromium in the environment. The occurrence of chromium in different water samples has been described in the paper. Moreover, the detection methods for the presence of chromium in these types of samples are also explained with the support of scientific papers and reports. Several advanced tools are available for the removal of chromium. However, more attempts are required to assess the presence of chromium. Furthermore, the standard guidelines should be adopted for the treatment of industrial and domestic effluent in India.

Acknowledgements The authors are thankful to GBPIHED, Kosi Katarmal, Almora for the support to complete this work and also thankful to the Director USERC for providing the necessary facilities.

References

Abdolmohammad-Zadeh H, Sadeghi GH (2012) A nano-structured material for reliable speciation of chromium and manganese in drinking waters, surface waters and industrial wastewater effluents. Talanta 94:201–208

Agrafioti E, Kalderis D, Diamadopoulos E (2014) Arsenic and chromium removal from water using biochars derived from rice husk, organic solid wastes and sewage sludge. J Environ Manage 133:309–314

Atieh MA (2011) Removal of Chromium (VI) from polluted water using carbon nanotubes supported with activated carbon. Procedia Environ Sci 4:281–293

ATSDR (2012) Toxicological profile for chromium. Agency for Toxic Substances and Disease Registry, Public Health Service, US Department of Health and Human Services

AWWA (American Water Works Association) (2013) Chromium in drinking water: a technical information primer. American Water Works Association (AWWA), Denver, Colorado
B'Hymer C, Brisbin JA, Sutton KL, Caruso JA (2000) New approaches for elemental speciation using plasma mass spectrometry. Am Lab 32:17–39
Background Technical Information for Hexavalent Chromium (Cr-6) (2015) Water research foundation
Bahadir Z, Bulut VN, Hidalgo M, Soylak M, Marguí E (2015) Determination of trace amounts of hexavalent chromium in drinking waters by dispersive microsolid-phase extraction using modified multi-walled carbon nanotubes combined with total reflection X-ray fluorescence spectrometry. Spectrochim Acta Part B: At Spectrosc 107:170–177
Beaumont JJ, Sedman RM, Reynolds SD, Sherman CD, Li L, Howd RA, Sandy MS, Zeise L, Alexeeff GV (2008) Cancer mortality in a Chinese population exposed to hexavalent chromium in drinking water. Epidemiology 19(1):12e23
Beller HR, Yang L, Varadharajan C, Ruyang H, Lim HC, Karaoz U, Molins S, Marcus MA, Brodie AL, Steefel CI, Nico PS (2014) Divergent aquifer biogeochemical systems converge on similar and unexpected Cr(VI) reduction products. Environ Sci Technol 48(18):10699e10706
Bhaumik M, Setshedi K, Maity A, Onyango MS (2013) Chromium (VI) removal from water using fixed bed column of polypyrrole/Fe_3O_4 nanocomposite. Sep Purif Technol 110:11–19
Campos J, Martinez-Pacheco M, Cervantes C (1995) Hexavalent-chromium reduction by a chromate-resistant Bacillus sp. strain. Antonie Van Leeuwenhoek 68(3):203–208
Catalani S, Fostinelli J, Enrica Gilberti M, Apostoli P (2015) Application of a metal free high performance liquid chromatography with inductively coupled plasma mass spectrometry (HPLC–ICP-MS) for the determination of chromium species in drinking and tap water. Int J Mass Spectrom 387:31–37
Charlet L, Manceau AA (1992) X-ray absorption spectroscopic study of the sorption of Cr(III) at the oxide-water interface, II. Adsorption, coprecipitation, and surface precipitation on hydrous ferric oxide. J Colloid Interface Sci 148(2):443e458
Chowdhury S, Jafar Mazumder MA, Al-Attas O, Husain T (2016) Heavy metals in drinking water: occurrences, implications, and future needs in developing countries. Sci Total Environ 569–570:476–488
Daugherty ML (1992) Toxicity summary for chromium. Oak Ridge National Laboratory Chemical Hazard Evaluation and Communication Group, Oak Ridge, TN
Eary LE, Rai D (1987) Kinetics of Chromium (III) oxidation to chromium (VI) by reaction with manganese dioxide. Environ Sci Technol 21:1187e1193
Ebdon L, Pitts L, Cornelis R, Crews H, Donard OFX, Quevauviller P (2001) Trace element speciation for environment, food and health. The Royal Society of Chemistry, Cambridge, UK
Guha H, Saiers JE, Brooks S, Jardine P, Jayachandran, K (2000) Chromium transport, oxidation, and adsorption in manganese-coated sand. Contam Hydrol 49(1e2):311e334
Gupta VK, Ali I (2004) Removal of lead and chromium from wastewater using bagasse fly ash—a sugar industry waste. J Colloid Interface Sci 271(2):321–328
Gupta VK, Nayak A, Agarwal S, Dobhal R, Uniyal DP, Singh P, Sharma B, Tyagi S, Singh R (2012) Arsenic speciation analysis and remediation techniques in drinking water. Desal Water Treat 40(1–3):231–243
Hackbarth FV, Maass D, de Souza AAU, Vilar VJP, de Souza SMAG (2016) Removal of hexavalent chromium from electroplating wastewaters using marine macroalga *Pelvetia canaliculata* as natural electron donor. Chem Eng J, 290:477–489
Harijan DKL, Chandra V (2016) Polyaniline functionalized graphene sheets for treatment of toxic hexavalent chromium. J Environ Chem Eng 4(3):3006–3012
Hem JD (1959) Study and interpretation of the chemical characteristics of natural water, 1st edn. U.S. Geological Survey Water-Supply, Paper 1473, 269 pp
IARC (International Agency for Research on Cancer) (2012) IARC Monographs (Volume 100C-9): a review of human carcinogens: arsenic, metals, fibres, and dusts: chromium (VI) compounds. International Agency for Research on Cancer (IARC)

Kaprara E, Kazakis N, Simeonidis K, Coles S, Zouboulis AI, Samaras P, Mitrakas M (2015) Occurrence of Cr(VI) in drinking water of Greece and relation to the geological background. J Hazard Mater 281:2–11
Khambhaty Y, Mody K, Basha S et al (2009) Kinetics equilibrium and thermodynamic studies on biosorption of hexavalent chromium by dead fungal biomass of marine Aspergillus niger. Chem Eng J 145:489–495
Korngold E, Belayev N, Aronov L (2003) Removal of chromates from drinking water by anion exchangers. Sep Purif Technol 33(2):179–187
Kotas J, Stasicka Z (2000) Chromium occurrence in the environment and methods of its speciation. Environ Pollut 107(3):263e283
Kumari M, Pittman CU Jr, Mohan D (2015) Heavy metals [chromium (VI) and lead (II)] removal from water using mesoporous magnetite (Fe_3O_4) nanospheres. J Colloid Interface Sci 442:120–132
Lee DC, Park CJ, Yang JE et al (2000) Screening of hexavalent chromium biosorbent from marine Algae. Applied Microbio Biotech 54:597–600
Mayer LM, Schick LL (1981) Removal of hexavalent chromium from estuarine waters by model substrates and natural sediments. Environ Sci Technol 15(12):1482e1484
Montes-Bayon M, DeNicola K, Caruso JA (2003) Liquid chromatography: inductively coupled plasma mass spectrometry. J Chromatogr A 1000:457–476
Motzer WE (2005) In: Guertin J, Jacobs JA, Avakian CP (eds) Chemistry, geochemistry, and geology of chromium and chromium compound, Chromium (VI) Handbook, 784 pp
Nourbaksh M, Sag Y, Ozer Z et al (1994) A comparative study of various biosorbents for removal Chromium (VI) ions from industrial wastewaters. Proce Biochem 29:1–5
Nriagu JO, Niebor E (1988) Chromium in the natural and human environments. Wiley, New York, 571 pp
Pérez E, Ayele L, Getachew G, Fetter G, Bosch P, Mayoral A, Díaz I (2015) Removal of chromium (VI) using nano-hydrotalcite/SiO_2 composite. J Environ Chem Eng 3(3):1555–1561
Perlmutter NM, Liber M, Frauenthal HL (1963) Movement of waterborne cadmium and hexavalent chromium wastes in South Farmingdale, Nassua County, Long Island. US Geological Survey Professional Paper 475C, pp C170eC184
Rai D, Zachara JM (1984) Chemical attenuation rates, coefficients, and constants in Leachate migration: volume 1: a critical review. Battelle Pacific Northwest Laboratories, Richland, WA. EA-3336, vol 1, Research Project 2198e1, variously paged
Reimann C, de Caritat P (1998) Chemical elements in the environment. Springer Verlag, Berlin, 398 pp
Robertson FN (1975) Hexavalent chromium in the ground water, in Paradise Valley, Arizona. Ground Water 13(6):516e527
Robertson FN (1991) Geochemistry of ground water in alluvial basins of Arizona and adjacent parts of Nevada, New Mexico, and California. US Geological Survey Professional Paper 1406-C, 90 pp
Sadyrbaeva TZ (2016) Removal of chromium (VI) from aqueous solutions using a novel hybrid liquid membrane—electrodialysis process. Chem Eng Process 99:183–191
Schroeder DC, Lee GF (1975) Potential transformations of chromium in natural waters. Water Air Soil Pollut 4:355e365
Sedman RM, Beaumont J, McDonald TA, Reynolds S, Krowech G, Howd R (2006a) Review of the evidence regarding carcinogenicity of hexavalent chromium in drinking water. J Environ Sci Health Part C 24(1):155e182
Sedman RM, Beaumont J, McDonald TA, Reynolds S, Krowech G, Howd R (2006b) Review of the evidence regarding the carcinogenicity of hexavalent chromium in drinking water. J Environ Sci Health Part C, Environ Carcinog Ecotoxicol 24:155–182
Selvi K, Pattabhi S, Kadirvelu K (2001) Removal of Cr(VI) from aqueous solution by adsorption onto activated carbon. Biores Technol 80(1):87–89

Silva B, Figueiredo H, Neves IC et al (2009) The role of pH on Cr(VI) reduction and removal by *Arthrobacter viscosus*. Int J Chem Bio Eng 2:100–103

Status of Trace and Toxic Metals in Indian Rivers (2014) River Data Directorate, Planning & Development Organisation, Central Water Commission, Ministry of Water Resources, Government of India, New Delhi

Stergioudi F, Kaprara E, Simeonidis K, Sagris D, Mitrakas M, Vourlias G, Michailidis, N (2015) Copper foams in water treatment technology: removal of hexavalent chromium. Mater Des 87:287–294

Ure AM (1990) Trace elements in soil: their determination and speciation. Fresenius J Anal Chem 337:567–581

U.S. Environmental Protection Agency (USEPA) (2008) Addressing chromium contamination in the San Fernando Valley. U.S. Environmental Protection Agency, Region 9, San Francisco, CA

WHO (World Health Organization) (2011) Guidelines for drinking-water quality, 4th edn. World Health Organization (WHO), Geneva, Switzerland

Yang L, You-zhao H, Wu-er G, Li Min, Qi-shu Q, Xiang-qin L (2001) Determination of chromium (VI) and lead(II) in drinking water by electrokinetic flow analysis system and graphite furnace atomic absorption spectrometry. Talanta 55(2):271–279

Yue Z, Bender SE, Wang J, Economy J (2009) Removal of chromium Cr(VI) by low-cost chemically activated carbon materials from water. J Hazard Mater 166(1):74–78

Ziagova M, Dimitriadis G, Aslanidou D et al (2007) Comparative study of Cd(II) and Cr(VI) biosorption on *Staphylococcus xylosus* and *Pseudomonas* Sp. In single and binary mixtures. Bio Resour Techn 98:2859–2865

'Phytowaste' and Vermicomposting as a Potential Route for Its Environmentally Compatible Utilization

S. A. Abbasi, Pratiksha Patnaik, Tabassum-Abbasi and Tasneem Abbasi

Abstract A new term 'phytowaste' has been introduced to denote that form of biomass or biowaste which comprises either the remains of the earlier live plant, or the present live plants which are weedy and are sought to be eradicated. The reasons why some species of plants, which are essential components of the ecosystems in their place of origin, become weedy when taken to new regions are enumerated. It is then shown that even as a lot of effort has been invested in either eradicating the weeds or finding ways to utilize them, neither have succeeded in controlling their spread. The prevailing means of utilizing most weeds are either not economical or are confined to their miniscule use in a few small regions of the world. In this context, the recently developed paradigm of 'high-rate vermicomposting' is discussed and its potential in converting billions of tonnes of 'phytowaste' into organic fertilizer is highlighted.

Keywords Biowaste · Phytomass · Invasives · High-rate vermicomposting · Organic fertilizer · Phytowaste

1 Introduction: How Phtomass Becomes a 'Waste'

1.1 Weeds as a Form of 'Phytowaste'

Generation of 'phytomass'—which is the biomass of botanical species or 'plants'—is the sole means of converting solar energy into a source of animal food; in other words, a means of primary production and primary consumption. If a species of phytomass in a region cannot perish at the rate at which it is generated, it can lead

S. A. Abbasi (✉) · P. Patnaik · T. Abbasi
Centre for Pollution Control and Environmental Engineering, Pondicherry University, Puducherry 605014, India
e-mail: abbbasi.cpee@gmail.com

Tabassum-Abbasi
Department of Health, Safety, Environment and Civil Engineering, University of Petroleum and Energy Studies, Dehradun 248007, India

N. A. Siddiqui et al. (eds.), *Advances in Water Pollution Monitoring and Control*, Springer Transactions in Civil and Environmental Engineering, https://doi.org/10.1007/978-981-32-9956-6_3

to a progressive excess of that species in that region. When such an excess begins to harm other species of plants in a location, outcompetes them, and monopolizes the use of soil, water and other natural resources of that location, it becomes a nuisance and is called a weed. A patch of land covered with a weed is similar in nature to a patch of land on which municipal solid waste has been dumped: harmful and undesirable.

But no species of plant is intrinsically harmful or worthless. Indeed, several species have been highly useful and essential in their regions of origin and have become serious pests only after they were introduced in other regions. Prosopis (*Prosopis juliflora*) is a major example. It has been considered a 'wonder tree' and perceived as a great blessing in its native desert environments (Patnaik 2017; Patnaik et al. 2017). But upon being introduced in other regions, it has become a very serious pest. Even olive tree, which is lovingly grown in most parts of the world, has become a weed in Australia as it has begun to invade and colonize lands which were not meant for it. *Hydrocotyle asiatica* is eaten as a vegetable in several regions of Asia while *Alternanthrea sessilis* and *Eclipta postrata* are keystone medicinal plants in the Ayurvedic and Homoeopathic systems of medicine (Swapna et al. 2011; Abhang et al. 2015; Jain et al. 2015; Himaja and Neelufar Shama 2015; Chung et al. 2017). Yet all the three are regarded as weeds because the extent of their growth is a lot more than that which is being utilized. Indeed, nearly every species presently recognized as a weed had been used in traditional medicine, or as animal feed, or as food, or as an ornamental plant (Abbasi and Abbasi 2010a, b; Abbasi et al. 2015). But the quantities thus utilized have been miniscule in comparison to the quantities generated.

1.2 Some Common Attributes of the Invasives

Not all exotic plants become weeds; those which become, have certain defining traits which include one or more of the following:

1. A very strong ability to reproduce and regenerate: Most invasive species have large seed banks with high viability. Most have the ability to regenerate sexually as well as vegetatively.
2. A high tolerance towards agroclimatic variations: Such plants can thrive in widely varying soil types, ambient humidity, ambient temperature, water availability, etc. They also trend to adapt to new environments quicker than other species do.
3. Great resilience and hardiness: They can withstand attempts at their eradication and keep coming back strongly.
4. Allelopathy and propensity to carry/release chemicals toxics to soil, animals (including humans) and other plants: This helps them elbow out other vegetation and hasten their colonization.
5. A pronounced ability to invade new regions and rapidly proliferate in them.
6. Presence of very few, if any, grazers or natural enemies: Plants which become invasive in exotic locales have enemies and grazers in their natural habitat which

keep their spread in check. When such plants are taken to regions where those natural enemies do not exist, the plants proliferate unchecked.

1.3 Aggravation Due to Anthropogenic Factors

The above-mentioned traits are aided and abetted by anthropogenic interventions such as (Abbasi and Abbasi 2018a, b)

1. Increasing conversion of natural forests into monocultures for commercial purposes
2. Environmental pollution
3. Interventions such as damming of rivers
4. Habitat fragmentation caused by unplanned and runaway 'development' in the form of roads, buildings, transmission lines, etc.

These and similar other forms of anthropogenic tampering of the environment reduces biodiversity and makes it increasingly difficult for the remaining of the sensitive species to survive. This paves the way for a few hardy and domineering species to invade more and more areas and colonize them.

1.4 The Impact

The impact of colonization of land or water by invasive plants (Fig. 1) is to worsen and hasten the eco-degradation already set in motion by the anthropogenic tampering with the environment. The impact manifests itself in the following ways (Abbasi and Nipaney 1986):

(i) Colonization of large areas by one or the other of these species which eliminates most other vegetation. With it are eliminated a large number of animals associated with those species of vegetation. It is also common to see 2–3 weeds like ipomoea, salvinia and water hyacinth pressing upon each other to gain accidency in an area.
(ii) There is increasing pressure on land and water resources, besides soil nutrients, which are monopolized by one or more of these weeds.
(iii) The habitat or food loss suffered by smaller animals—who had depended on the vegetation since repelled by these weeds—adversely affects the animals higher in the food chain. It also favours a small number of hardy animals capable of surviving in the weed monocultures, allowing them to proliferate and dominate at the cost of more sensitive and niche-specific animals, thereby harming the biodiversity even further.
(iv) Serious jeopardies to the aesthetics of the water—spreads, water holding capacity of the lakes and reservoirs, water quality, fisheries, navigation, water sports,

Fig. 1 Colonization of land by ipomoea (top, left), lantana (top, right) and parthenium (below, left). A canal having been so overwhelmed by water hyancith (below, right) that water has been pushed out of sight

etc.—are caused when the invasives happen to be an aquatic weed like salivinia (*Salvinia molesta*), or an amphibious weed like ipomoea (*Ipomoea carnea*).

The cumulative losses caused by these weeds in terms of loss of forest and arable land, water quality and quantity, soil nutrients, biodiversity, ecosystem services and habitats are estimated to run into billions of rupees per year.

1.5 Weeds as Forms of Phytowaste

The phytomass represented by weeds can be regarded as 'phytowaste' for the following reasons:

(i) It does not perform ecosystem services as numerous other species (which are not directly used by humans but indirectly contribute to ecosystem health) do. On the other hand, it interferes with and harms those species which perform ecosystem services.

(ii) It either has no direct utility at all or has much lesser utility in comparison to its availability.
(iii) By eliminating or discouraging most sensitive species of plants and animals, it facilitates the proliferation of a few hardy species which can be disease vectors. For example, salvinia provides habitat for mosquitoes which can cause malaria, dengue and elephantiasis. It also facilitates the species of snail which cause schistomiasis.
(iv) When it falls on soil/water in the form of dead leaves/twigs or dead whole plants, it degrades aerobically or anaerobically—mostly latter. The former generates CO_2 while the latter emits a 1:3 mixture of CO_2 and CH_4 (methane). Given that each molecule of CH_4 causes 25 times more global warming than each molecule of CO_2 (34 times by the estimate of Shindell and Faluvegi 2009), this type of degradation in nature amounts to a plant contributing more global warming gas (GWG) than it had earlier fixed (in the form of atmospheric or aquatic CO_2) in the course of photosynthesis.

Viewed from this perspective, stands of weeds like salvinia, ipomoea or lantana are veritable 'waste factories' because they keep spreading and consolidating thereby contributing more and more 'waste' phytomass. Salvinia, alongside water hyacinth (*Echhornia crassipes*), is the most productive of all plants, attaining net primary production of the order of 60 ash-free dry tonnes per hectare, per year (Abbasi and Nipaney 1995).

1.6 Huge Streams of Other Phytowaste

Massive streams of phytowaste are also generated in the course of agriculture, especially horticulture. An example is the biodegradable waste generated to the tune of 105 million tonnes per year (Al-Juhaimi et al. 2014) in the course of cultivation of date palm trees (*Phoenix dactylifera*). As is the case with the weed phytomass, these streams are also largely unutilized at present and degrade in the open, releasing a more potent green house gas—CH_4—than the one they had fixed earlier, viz. CO_2.

2 Past Efforts to Eradicate or Control the Weeds

Classically, infestation by any plant is controlled by any one or the combinations of the following means:

(i) Physical removal
(ii) Killing by chemicals
(iii) Biological control.

Since over a hundred years, especially during the second half of the twentieth century, very strong efforts have been made all over the world to find ways to destroy unwanted plants (Abbasi and Nipaney 1993b). Indeed, minor weeds can be, and are, controlled by periodic manual removal and a large number of chemicals have been tried with which to kill seeds and seedlings of the weeds. In like manner, biological methods, mainly revolving round weed-specific grazers, have been tried in profusion. But no major invasive has even been controlled, let alone destroyed till now. Once in a while, a temporary subduing of an infestation is achieved but either the targeted invasive itself stages a comeback or some other equally pernicious invasive takes over.

Worse, chemical and biological methods carry the grave risk of harming non-target species and also toxifying the environment. Moreover, a biological control agent can itself go out of control and become a pest itself.

For all these reasons, invasive plants, as a group, have not only withstood all attempts at controlling them but they are also colonizing more and more areas with time.

3 Impact of Infestation by Weeds as Illustrated with the Examples of Salvinia and Ipomoea

Salvinia (*Salvinia molesta,* D. S. Mitchell) is an exceedingly invasive and dominant aquatic weed, capable of multiplying and growing faster than most other known botanical species.

It has occupied large areas of wetlands in South America, Africa, South Asia and Australia (Abbasi and Nipaney 1993b; Hussain et al. 2016a), resulting in the generation of billions of tonnes of salvinia biomass every year. Since no process exists to utilize any significant proportion of the salvinia biomass (Hussain et al. 2018), it remains unharvested, causing serious harm to the biodiversity and health of the wetlands (Abbasi and Nipaney 1993b).

Indeed due to its highly invasive and colonizing attributes, salvinia has been included in the list of '100 of the world's worst invasive alien species' (Luque et al. 2013; GISD 2017). This has happened after the International Union for Conservation of Nature (IUCN) that assessed more than 10,000 invasive species from the world's largest databases for their capacity to spread and for their potential ecological or economic impact. According to Luque et al. (2013), 'more than 650 experts from 63 countries then voted on the ten candidate species that were shortlisted, and selected the giant salvinia (*Salvinia molesta*)…native to Brazil, this fern has spread throughout the tropics and subtropics. It doubles in abundance within days, forming thick, floating mats that block light from expanses of water, reduce its oxygen content and degrade water quality. They also impede water-based transport, clog irrigation and power generation systems, and harm local fisheries. Now in the global spotlight, this new entrant to the IUCN list is set to increase public awareness of the harm caused by invasive species and to stimulate more discussion in science and policy circles'.

Once in a while, success is achieved in halting the rapidly growing spreads of salvinia by the use of biological agents (Room et al. 1981) but such successes have been rare and have not been replicated. The same biocontrol agent which might have been effective in a particular situation has been found wanting in other situations (Abbasi and Nipaney 1993a). Chemical and mechanical methods have seen even lesser success in controlling salvinia.

Ipomoea (*Ipomoea carnea* Jacq., also called *Ipomoea fistulosa*) is an evergreen, perennial, fast-growing, amphibious shrub. It attains heights ranging from 1.1 to 3 m and stem diameter between 1.5 and 6 cm. It was initially used to make fences—its violet flowers being an attraction—but has since, metaphorically, crossed all fences to invade and colonize landmasses and wetlands everywhere. Ipomoea is able to adapt to very diverse habitats and environmental conditions. This, coupled by its high productivity and hardiness, has made it into one of the most dominant and harmful of the world's weeds (Shaltout et al. 2010; Kumar et al. 2014, 2015).

Ipomoea contains several alkaloids, which possess mammalian toxicity. It can cause chronic neurologic disease characterized by weight loss, depression, altered behaviour, infertility, birth defects and death of animals which graze on its leaves (Panter et al. 1999; Hueza et al. 2003; Almeida Rios et al. 2009; Cook et al. 2015). It also contains allelopathic compounds, which repel or toxify other vegetation (Kumar et al. 2018).

4 Attempts at Utilization of the Common Weeds

There have been concerted attempts to find ways and means of utilizing weeds and other phytowaste, including salvinia and ipomoea. These aspects have been reviewed by us recently in respect of ipomoea, lantana, parthenium and salvinia (Kumar et al. 2018). The studies reveal that the attempts to utilize these weeds encompass a wide range and include utilization as

- Antimicrobial agents
- Insecticides, helminthicides, fungicides
- Source of drugs
- Source of other useful chemicals
- Source of activated carbon
- Feedstock in pyrolysis and gasification plants
- As biosorbents (other than as activated carbon)
- Source of biofuels
- Agents for biomimetic nanoparticle synthesis
- Source of heat transfer fluids
- Agents for phytoremediation
- Agents for phytoextraction
- Corrosion inhibitors
- Source of antioxidants

- Compost
- Source of energy precursors in the form of volatile fatty acids
- Source of fermentable sugars
- Paper pulp
- Additive to cow-dung in generating vermicompost.

But, none of these methods have proved economically viable so far. Second, even if one or other of the above-mentioned options become viable, the extent to which the weeds can be used for that purpose will be very small compared to the quantities in which these weeds are generated every year. As stated in Sect. 1.1, some of the weeds are keystone medicines, and some are consumed as vegetables in a few regions of the world. But the quantities in which these weeds are utilized are insignificant in comparison to the quantities that are generated.

5 The Potential of Vermicomposting

As brought out by Abbasi et al. (2015), to quote, 'earthworms process enormous quantities of leaf litter and other forms of plant debris in nature by ingesting them and converting them into vermicast. The latter is widely recognized as a soil-friendly and plant-friendly organic fertilizer. Earthworms also consume animal droppings but the quantities of plant biomass processed by earthworms are several times greater than the quantities of zoomass they handle. Yet, when controlled vermicomposting is done to process biodegradable solid waste, terrestrial/aquatic weeds, or crop waste is rarely vermicomposted on a large scale'. Unquote.

As noted earlier, weeds like water hyacinth, salvinia, ipomoea, lantana and parthenium generate billions of tonnes of biomass per year. This ever-increasing biomass remains unutilized. Besides harming the environment in many ways, this also contributes to global warming (Abbasi and Abbasi 2011) as the debris and dead plants of the weeds degrade in the open, generating CO_2 or CO_2–CH_4 mixtures depending on whether the degradation occurs aerobically or anaerobically. Given this context, developing an inexpensive and clean process with which huge quantities of the weed biomass can be profitably utilized is a major challenge (Abbasi and Abbasi 2010a, b; Abbasi et al. 2015).

6 The Reasons Why the Potential of Generating Fertilizers from the Vermicomposting of Phytomass Was Unutilized So Far

We have recently enumerated (Abbasi et al. 2015) the reasons why vermicomposting is arguably the best option for the utilization of biodegradable solid waste, especially phytomass:

(a) As noted by Abbasi et al. (2015), to quote, 'There is a rapidly growing interest in the vermicomposting of waste phytomass, especially since the turn of the present century. It is perhaps due to the increasing appreciation that vermicomposting is a phytomass utilization option which can generate good quality organic fertilizer. There is also an increasing realization that other options of phytomass utilization, such as composting and anaerobic digestion are not only more cumbersome and expensive but incapable of handling phytomass (Abbasi et al. 2012a, b). Sanitary landfills, which at present handle larger volumes of biodegradable solid waste than any other process does, are already overburdened, besides being inherently unsuitable for phytomass (Abbasi 2018). Moreover, at best only 60% of the methane that is generated by a landfill—often much lesser or none e can be captured while the rest gets released into the atmosphere (Ritzkowski and Stegmann 2007; Zamorano et al. 2007). These aspects put sanitary landfills among of the world's major sources of global warming gases (Abbasi et al. 2012a). This is more so because each molecule of methane contributes 34 times to the global warming as compared to a molecule of carbon dioxide (Shindell and Faluvegi 2009)'. Unquote.

(b) In contrast, as per Abbasi et al. (2015), to quote 'vermicomposting is an aerobic process and only about 40% of the carbon contained in the phytomass is released as CO_2. The rest is returned to the soil as vermicompost. As the CO_2 released from the phytomass comes from carbon that had already been sequestered, vermicomposting results in additional carbon sequestration'. Unquote.

In this back-drop, Abbasi et al. (2015) have identified the reasons why past attempts at vermicomposting of phytomass have been unviable. All of them are attributable to the intrinsic weakness of the hitherto used batch-fed vermireactors due to the low surface area-height ratios of such reactors and other lacunae in design and operation. In the efforts to overcome this limitation, animal manure is added to phytomass in 1:1 mass ratio and/or pre-composting of the phytomass is done for achieving its vermicomposting. According to Abbasi et al. (2015), quote 'this makes the overall vermicomposting process more cumbersome and time-consuming, hence potentially costlier, than direct vermicomposting. It also drastically limits the quantities of phytomass that can be utilized as vermireactor feed because much lesser quantities of animal manure are available for proportionate supplementation'. Unquote.

More specifically, conventional vermicomposting technology is besieged with the following problems:

(a) *Slowness of the existing process designs*: They take 4–6 months for the input feed to be quantitatively converted to vermicast.

(b) *Heavy reliance on animal manure*: The substrate traditionally used for vermicomposting has been animal manure. Several authors have used other substrates like vegetable waste, garden trimmings and waste paper as vermireactor feed but always as a supplement to animal manure. But animal manure has several other remunerative uses, especially in developing countries, and is also a preferred feed for anaerobic digesters. If substrates other than animal manure have to be brought within the scope of vermicomposting, the reliance of vermicomposting on animal manure must be drastically reduced so that other types of feedstock can be employed in vermireactors with little or no necessity of blending it with animal manure.

(c) *Time–consuming nature of pre-composting*: Most pre-existing processes for vermicomposting plant biomass require pre-composting. This adds time and cost to the overall process, affecting the economy of the process.

(d) *Hazards in the collection and transport of animal manure*: This leads to the emission of global warming gases and nitrous oxide (Tauseef et al. 2013) besides other pollutants like ammonia.

7 The Paradigm of High-Rate Vermicomposting and Its Essential Features

In this backdrop, S. A. Abbasi and coworkers have developed the paradigm of high-rate vermicomposting and associated technology (Abbasi et al. 2009, 2011; Tauseef et al. 2013, 2014) which enables direct and efficient vermicomposting of phytomass. According to Abbasi et al. (2015), the special attributes of 'high-rate vermicomposting' paradigm, which enables direct vermicomposting of phytomass, are, to quote '(a) it relies on a reactor geometry that has been chosen to maximize earthworm-substrate contact as well as ease of cast deposition. (b) To this end, a high surface area-to-volume ratio is set for the reactors. (c) This also ensures mixing and aeration of the reactor content by the earthworm movements thereby preventing anaerobic pockets from developing. (d) The vermicast harvesting is also made easier by the low aspect-ratio because many species deposit their cost on top of the substrate while some others do it at the bottom. (e) Low substrate column height makes it possible to maintain, with relative ease, uniformity in the moisture content across the reactor depth. There is little, if any, accumulation of water in the reactor bottom, saving on the need for recycling. (f) The sand-gravel "vermibed", which occupies over 25% space in conventional vermireactors, is replaced by moist jute cloth. (g) This maximizes the use of reactor volume, proportionately reducing the system cost. (h) Earthworm density is maximized to achieve highest sustainable population for a given feed. (i) The high earthworm: feed ratio further helps mixing and aeration of the substrate due to earthworm movement'. Unquote.

Abbasi and coworkers have also demonstrated the success of their technology by direct and efficient vermicomposting of a large number of phytomass species (Tauseef et al. 2014; Nayeem-Shah et al. 2015), besides paper waste. These have proved to be as effective fertilizers (Fig. 2) as manure-based vermicompost is known to be (Hussain et al. 2015, 2016a, b, 2017; Hussain and Abbasi 2018). Volatile fatty acids (VFAs) can be extracted from the phytomass by soaking the phytomass in cowdung-inoculated water before subjecting the phytomass to vermicomposting. The VFAs thus extracted are utilizable as energy precursors (Ganesh et al. 2005; Kumar et al. 2015). In the like manner biomolecules can be extracted in water from

Fig. 2 Tomatos (left) and brinjal cultivated using vermicompost of the leaves of prosopis (*Prosopis juliflora*)

the phytomass for generating nanoparticles (Anuradha et al. 2015a, b; Mavukkandy et al. 2016).

8 Summary

The paper begins with the exposition that the biomass generated by high colonizing of invasive botanical species (weeds) is essentially a waste. It has the same types of harmful effects on the environment as on other forms of biowaste. Accordingly, for such phytomass, the authors have proposed the name 'phytowaste'. The paper then reviews the past attempts to eradicate or control the invasives and shows that none of the attempts has succeeded in controlling any invasive except locally and for short durations. Worse, as soon as the hold of one colonized species is weakened, other invasives move in. The past attempts to utilize phytowaste have also fallen short of their promise partly due to economic unviability and partly due to their inability to utilize significant quantities of the phytowaste. Given this backdrop, the authors have argued that vermicomposting is an option that is likely to be not only economically viable but capable of generating a product—organic fertilizer—that has unlimited demand in the market. The authors recount why this potential of vermicomposting has been untapped till now and report the recent breakthroughs that promise to remedy these situations. The breakthroughs are built around the paradigm of 'high-rate vermicomposting' and associated technology developed by the authors. The distinguishing features of the new technology have been recounted.

Acknowledgements SAA thanks the Council of Scientific and Industrial Research (CSIR), New Delhi, for the Emeritus Scientist grant (21(1034)/16/EMR-II).

References

Abbasi SA (2018) The myth and the reality of energy recovery from municipal solid waste. Energy Sustain Soc 8(1)

Abbasi T, Abbasi SA (2010a) Renewable energy sources: their Impact on global warming and pollution. PHI Larning, New Delhi, pp x + 306

Abbasi T, Abbasi SA (2010b) Production of clean energy by anaerobic digestion of phytomass—new prospects for a global warming amelioration technology. Renew Sustain Energy Rev 14(6):1653–1659

Abbasi T, Abbasi SA (2011) Decarbonization of fossil fuels as a strategy to control global warming. Renew Sustain Energy Rev 15(4):1828–1834

Abbasi T, Abbasi SA (2018a) Perspectives in pollution control and sustainable development. Discovery Publishing House, New Delhi, xi+569 pp. ISBN: 978-9-3505-6889-7

Abbasi SA, Abbasi T (2018b) Current concerns in environmental engineering. Nova Science Publishers, New York, pp xvi+309. ISBN: 978-1-53613-920-4

Abbasi SA, Nipaney PC (1984) Generation of biogas from *Salvinia molesta* (Mitchell) on a commercial biogas digester. Environ Technol 5(1–11):75–80

Abbasi SA, Nipaney PC (1985) Wastewater treatment using aquatic plants. Survivability and growth of *Salvinia molesta* (Mitchell) over waters treated with zinc (II) and the subsequent utilization of the harvested weeds for energy (biogas) production. Resour Conserv 12(1):47–55

Abbasi SA, Nipaney PC (1993a) Modelling and simulation of biogas systems economics. Ashish Publishing House, New Delhi, xviii + 356 pp

Abbasi SA and Nipaney PC (1993b) World's worst weed: impact and control. International Book Distributors, Dehradun, xi + 226 pp

Abbasi PC, Nipaney (1986) Infestation by aquatic weeds of the fern genus: its status and control. Environ Conserv 13(3):235–241

Abbasi SA, Nipaney PC (1994) Potential of aquatic weed Salvinia molesta (Mitchell) for water treatment and energy recovery

Abbasi SA, Nipaney PC (1995) Productivity of aquatic weed Salvinia (*Salvinia molesta,* Mitchell) in natural waters. Ecol Environ Conserv 1(1–4):11–12

Abbasi SA, Ramasamy EV (2001) Solid waste management with vermitechnology, vol xiv. Discovery Publishing House, New Delhi, p 178

Abbasi SA, Nipaney PC, Schaumberg GD (1990) Bioenergy potential of eight common aquatic weeds. Biol Wastes 34(4):359–366

Abbasi T, Chari KB, Abbasi SA (2008) Oussudu lake, Pondicherry, India: a survey on socio-economic interferences. Indian Geogr J 83(2):149–162

Abbasi T, Gajalakshmi S, Abbasi SA (2009) Towards modeling and design of vermicomposting systems: mechanisms of composting/vermicomposting and their implications. Indian J Biotechnol 8:177–182

Abbasi T, Tauseef SM, Abbasi SA (2011) The inclined parallel stack continuously operable vermireactor. Off J Pat Off 22:9571

Abbasi T, Tauseef SM, Abbasi SA (2012a). Biogas energy. Springer, New York & London, xiv+169 pp. ISBN 978-1-4614-1039-3

Abbasi T, Tauseef SM, Abbasi SA (2012b) Anaerobic digestion for global warming control and energy generation: an overview. Renew Sustain Energy Rev 16(5):3228–3242

Abbasi SA, Nayeem-Shah M, Abbasi T (2015) Vermicomposting of phytomass: limitations of the past approaches and the emerging directions. J Clean Prod 93:103–114

Abdul Syukor bin AR, Sulaiman SB (2014) Treatment of i{dotless}ndustrial wastewater using Eichornia crassipes, Pistia Stratiotes and Salvinia Molesta in phytogreen system. Energy Educ Sci Technol Part A: Energy Sci Res 32(1):339–346
Abhang AR, Pathare SA, Rohokale PG (2015) Traditional uses of medicinal plants by tribal and rural folk from Mula river valley of Rahuri and its adjoining area of Parner and Sangamner tahsils of Ahmednagar District (M.S.). Int J Pharma Bio Sci 6(1):P140–P148
Ahmad J, Abdullah SRS, Hassan HA, Rahman RAA, Idris M (2017) Screening of tropical native aquatic plants for polishing pulp and paper mill final effluent. Malaysian J Analytical Sci 21(1):105–112
Al-Juhaimi FY, Hamad SH, Al-Ahaideb S, Al-Otaibi MM, Ghafoor K, Abbasi T, Abbasi SA (2014) Biogas production through the anaerobic digestion of date palm tree waste-process optimization. Bioresources 9:3323–3333
Almeida Rios SD, Paes D, Borém A (2009) Adaptability and stability of carotenoids in maize cultivars. Crop Breed Appl Biotechnol 9(4):313–319
Anuradha J, Abbasi T, Abbasi SA (2015a) An eco-friendly method of synthesizing gold nanoparticles using an otherwise worthless weed pistia (*Pistia stratiotes* L.) J Adv Res 6(5):711–720
Anuradha J, Abbasi T, Ganaie SU, Abbasi SA (2015b) Gainful utilization of the highly intransigent weed ipomoea in the synthesis of gold nanoparticles. J King Saud Univ-Sci (Elsevier) 27:15–22
Arthur GD, Stirk WA, Novák O, Hekera P, Van Staden J (2007) Occurrence of nutrients and plant hormones (cytokinins and IAA) in the water fern *Salvina molesta* during growth and composting. Environ Exp Bot 61(2):137–144
Ashraf MA, Maah MJ, Yusoff I, Gharibreza MM (2010) Heavy metals accumulation and tolerance in plants growing on ex-mining area, Bestari Jaya, Kuala Selangor, Peninsular Malaysia. In: 2010 International Conference on Environmental Engineering and Applications (ICEEA), pp 267–271
Ashraf MA, Maah MJ, Yusoff I (2011) Heavy metals accumulation in plants growing in ex tin mining catchment. Int J Environ Sci Technol 8(2):401–416
Ashraf MA, Maah MJ, Yusoff I (2012) Assessment of phytoextraction efficiency of naturally grown plant species at the former tin mining catchment. Fresenius Environ Bull 21(3):523–533
Chandanshive VV, Rane NR, Gholave AR, Patil SM, Jeon BH, Govindwar SP (2016) Efficient decolorization and detoxification of textile industry effluent by *Salvina molesta* in lagoon treatment. Environ Res 150:88–96
Chantiratikul P, Meechai P, Nakbanpotec W (2009) Antioxidant activities and phenolic contents of extracts from *Salvinia molesta* and Eichornia crassipes. Res J Biol Sci 4(10):1113–1117
Chari ST, Leibson CL, Rabe KG, Ransom J, De Andrade M, Petersen GM (2005) Probability of pancreatic cancer following diabetes: a population-based study. Gastroenterology 129(2):504–511
Choudhary MI, Naheed N, Abbaskhan A, Musharraf SG, Siddiqui H (2008) Phenolic and other constituents of fresh water fern *Salvinia molesta*. Phytochemistry 69(4):1018–1023
Chung I, Rahuman AA, Marimuthu S, Kirthi AV, Anbarasan K, Padmini P, Rajakumar G (2017) Green synthesis of copper nanoparticles using eclipta prostrata leaves extract and their antioxidant and cytotoxic activities. Exp Ther Med 14(1):18–24. https://doi.org/10.3892/etm.2017.4466
Cook BI, Ault TR, Smerdon JE (2015) Unprecedented 21st century drought risk in the American Southwest and Central Plains. https://doi.org/10.1126/sciadv.1400082
Crites RW (2006) Natural wastewater treatment systems. CRC Press, Boca Raton, FL
Dwiloka B, Setiadi A, Santoso SI, Suprijatna E, Susanti S (2015) Effects of duck feed supplemented with invasive giant salvinia (*Salvina molesta*) on duck meat characteristics. Turkish J Vet Anim Sci 39(6):668–675
Edwards CA, Norman QA, Sherman R (2011) Vermiculture technology, earthworms, organic waste and environmental management. CRC Press, pp 17–19
Espinoza-Quinones FR, Zacarkim CE, Palacio SM, Obregon CL, Zenatti DC, Galante RM, Rossi N, Rossi FL, Pereira IRA, Welter RA, Rizzutto MA (2005) Removal of heavy metal from polluted river water using aquatic macrophytes Salvinia sp. Braz J Phys 35(3B):744–746

Finlayson CM, Farrell TP, Griffiths DJ (1984) Studies of the hydrobiology of a tropical lake in north-western Queensland. III. Growth, chemical composition and potential for harvesting of the aquatic vegetation. Mar Freshw Res 35(5):525–536

Ganesh PS, Ramasamy EV, Gajalakshmi S, Abbasi SA (2005) Extraction of volatile fatty acids (VFAs) from water hyacinth using inexpensive contraptions, and the use of the VFAs as feed supplement in conventional biogas digesters with concomitant final disposal of water hyacinth as vermicompost. Biochem Eng J 27(1):17–23

GISD (2017) http://www.iucngisd.org/gisd/search.php. Accessed 13 Jan 2017

Gupta M, Devi S (1995) Uptake and toxicity of cadmium in aquatic ferns. J Environ Biol 16(2):131–136

Henares MNP, Camargo AFM (2014) Estimating nitrogen and phosphorus saturation point for Eichhornia crassipes (Mart.) Solms and *Salvinia molesta* Mitchell in mesocosms used to treating aquaculture effluent. Acta Limnologica Brasiliensia 26(4):420–428

Henares MNP, Camargo AFM (2014b) Treatment efficiency of effluent prawn culture by wetland with floating aquatic macrophytes arranged in series. Braz J Biol 74(4):906–912

Henry-Silva GG, Camargo AFM (2006) Efficiency of aquatic macrophytes to treat Nile tilapia pond effluents. Scientia Agricola 63(5):433–438

Henry-Silva GG, Monteiro CAF (2002) Nutritive value of free-floating aquatic macrophytes (*Eichhornia crassipes, Pistia stratiotes and Salvinia molesta*), used in aquaculture waste treatment. Acta Sci-Biol Health Sci 24(2):519–526

Himaja N, Neelufar Shama S (2015) Herbal wealth for hepatotoxicity: a review. Asian J Pharm Clin Res 8(1):3–9

Hueza IM, Dagli ML, Gorniak SL, Paulino CA (2003) Toxic effect of pre-natal Ipomoea carnea administration to rats. Vet Hum Toxicol 45:298–302

Hunt J, Bhushan B (2011) Nanoscale biomimetics studies of *Salvinia molesta* for micropattern fabrication. J Colloid Interface Sci 363(1):187–192

Hussain N, Abbasi SA (2018) Efficacy of the vermicomposts of different organic wastes as "clean" fertilizers: state-of-the-art. Sustainability 10(4):1205

Hussain N, Abbasi T, Abbasi SA (2015) Vermicomposting eliminates the toxicity of Lantana (*Lantana camara*) and turns it into a plant friendly organic fertilizer. J Hazard Mater 298:46–57

Hussain N, Abbasi T, Abbasi SA (2016a) Vermiremediation of an invasive and pernicious weed salvinia (Salvinia molesta). Ecol Eng 91:432–440

Hussain N, Abbasi T, Abbasi SA (2016b) Transformation of toxic and allelopathic lantana into a bening organic fertilizer through vermicomposting. Spectrochim Acta Part A: Mol Biomol Spectrosc 163:162–169

Hussain N, Abbasi T, Abbasi SA (2017) Toxic and allelopathic ipomoea yields plant-friendly organic fertilizer. J Clean Prod 148:826–835

Hussain N, Tauseef SM, Abbasi SA (2018) Generation of highly potent organic fertilizer from pernicious aquatic weed *Salvinia molesta*. Environ Sci Pollut Res 25:4989–5002

Jain S, Choudhary GP, Jain DK (2015) Medicinal plants with potential anti-fertility activity: a review. Int J Green Pharm 9(4):223–228

Khan E, Virojnagud W, Ratpukdi T (2004) Use of biomass sorbents for oil removal from gas station runoff. Chemosphere 57(7):681–689

King C, McIntosh D, Fitzsimmons K (2004) Giant Salvinia (*Salvinia molesta*) as a partial feed for Nile Tilapia (Oreochromis niloticus). In: Proceedings of 6th International Symposium Tilapia in Agric, September, pp 12–16

Kumar S, Pandey S, Pandey AK (2014) In vitro antibacterial, antioxidant, and cytotoxic activities of Parthenium hysterophorus and characterization of extracts by LC-MS analysis. BioMed Res Int

Kumar MR, Tauseef SM, Abbasi T, Abbasi SA (2015) Control of amphibious weed ipomoea (Ipomoea carnea) by utilizing it for the extraction of volatile fatty acids as energy precursors. J Adv Res 6(1):73–78

Kumar MR, Tauseef SM, Abbasi T, Abbasi SA (2018) Invasiveness and colonizing ability of ipomoea (Ipomoea carnea) and attempts at its management. Nat Environ Pollut Technol 17(3):767–775

Lakra KC, Lal B, Banerjee TK (2017) Decontamination of coal mine effluent generated at the Rajrappa coal mine using phytoremediation technology. Int J Phytorem 19(6):530–536

Li S, Wang P, Deng G, Yuan W, Su Z (2013) Cytotoxic compounds from invasive giant salvinia (*Salvina molesta*) against human tumor cells. Bioorg Med Chem Lett 23(24):6682–6687

Luque AE, Corales R, Fowler RJ, Dimarco J, van Keken A, Winters P, Keefer MC, Fiscella K (2013) Bridging the digital divide in HIV care: a pilot study of an iPod personal health record. J Int Assoc Physicians AIDS Care 12:117–121. https://doi.org/10.1177/1545109712457712

Mavukkandy MO, Chakraborty S, Abbasi T, Abbasi SA (2016) A clean-green synthesis of platinum nanoparticles utilizing a pernicious weed Lantana (*Lantana Camara*). Am J Eng Allied Sci 9(1):84–90

Matamoros V, Nguyen LX, Arias CA, Salvadó V, Brix H (2012) Evaluation of aquatic plants for removing polar micro contaminants: a microcosm experiment. Chemosphere 88(10):1257–1264

Mishra R, Dwivedi S, Misra S (2010) Chromium removal from tannery effluent by phytoremediation. Pollut Res 29:69–71

Mithraja MJ, Marimuthu J, Mahesh M, Paul ZM, Jeeva S (2011) Phytochemical studies on Azollapinnata R. Br., Marsileaminuta L. and *Salvina molesta* Mitch. Asian Pac J Trop Biomed 1(1):S26–S29

Mohanty PK, Mishra D (1963) Stomatal distribution in relation to xeromorphy in aquatic plants. Nature 200(4909):909–910

Moozhiyil M, Pallauf J (1986) Chemical composition of the water fern, *Salvinia molesta*, and its potential as feed source for ruminants. Econ Bot 40(3):375–383

Mubarak M, Shaija A, Suchithra TV (2016) Optimization of lipid extraction from *Salvina molesta*. Environ Sci Pollut Res 23(14):14047–14055

Nayeem-Shah M, Gajalakshmi S, Abbasi SA (2015) Direct, rapid and sustainable vermicomposting of the leaf litter of neem (*Azadirachta indica*). Appl Biochem Biotechnol 2(175):792–801

Ng YS, Chan DJC (2017) Wastewater phytoremediation by *Salvina molesta*. J Water Process Eng 15:107–115

Ng YS, Samsudin NIS, Chan DJC (2017) Phytoremediation capabilities of *Spirodela polyrhiza* and *Salvina molesta* in fish farm wastewater: a preliminary study

Nitha TG, Jayanthi J, Ragunathan MG (2016) Antioxidant activity, total phenol, flavonoid, alkaloid, tannin, and saponin contents of leaf extracts of *Salvina molesta*. Asian J Pharm Clin Res 9(1):200–203

Nithya TG, Jayanthi J, Raghunathan MG (2015) Phytochemical, antibacterial and GC MS analysis of a floating fern *Salvina molesta* DS Mitchell (1972). Int J PharmTech Res 8(9):85–90

Panter KE, James, LF, Gardner, DR (1999) Lupines, poison–hemlock and *Nicotiana* spp: toxicity and teratogenicity in livestock. J Nat Toxins 8:117–134

Patnaik P (2017) Towards a clean technology for the total utilization of the invasive xerophyte prosopis (*prosopis juliflora*, swartz, dc.). Pondicherry University thesis

Patnaik P, Abbasi T, Abbasi SA (2017) Prosopis (*Prosopis juliflora*): blessing and bane. Trop Ecol in press

Raju RA, Gangwar B (2004) Utilization of potassium-rich green-leaf manures for rice (Oryza sativa) nursery and their effect on crop productivity. Indian J Agron 49(4):244–247

Ranjitha J, Raj A, Kashyap R, Vijayalakshmi S, Donatus M (2016) Removal of heavy metals from industrial effluent using Salviniamolesta. Int J ChemTech Res 9(05):608–613

Razak ASBA, bin Ab Wahid Z, bin Zakaria I, Said MIBM (2013) Treatment of industrial wastewater at Gebeng area using *Eichornia Crassipes* Sp. (Water Hyacinth), *Pistia Stratiotes* Sp. (Water Lettuce) and *Salvina molesta* Sp. (Giant Salvinia). Adv Environ Biol 7(12), 3802–3808

Ribeiro TH, Smith RW, Rubio J (2000) Sorption of oils by the nonliving biomass of a Salvinia sp. Environ Sci Technol 34(24):5201–5205

Ritzkowski M, Stegmann R (2007) Controlling greenhouse gas emissions through landfill in situ aeration. Int J Greenhouse Gas Control 1(3):281–288

Room PM, Harley KLS, Forno IW, Sands DPA (1981) Successful biological control of the floating weed salvinia. Nature 294(5836):78–80

Sandhyasree M, Kruthika DL, Priyanka B, Vijayalakshmi S, Ranjitha, J (2015) Microwave-assisted digestion for determination of Pb, Mg, Mn, Cd and Zn in *Salvina molesta* by flame atomic absorption spectrometry. International Journal of Pharmaceutical Sciences Review and Research 32:95–97

Santoso SI, Setiadi A (2016) Profitable utilization of Giant Salvinia, *Salvina molesta*, as local duck feed. Int J Pharm Sci Rev Res 15(4):121

Shaltout KH, Al-Sodany YM, Eid EM (2010) Growth behaviour of the invasive species Ipomoea carnea in the Nile Delta, Egypt. Hydrobiologia 656(1):187–197

Shindell D, Faluvegi G (2009) Climate response to regional radiative forcing during the twentieth century. Nat Geosci 2(4):294

Shiny KJ, Remani KN, Jalaja TK, Sasidharan VK (2004) Removal of chromium by two aquatic pteridophytes. J Environ Sci Eng 46(3):249–251

Singh D, Suthar S (2012) Vermicomposting of herbal pharmaceutical industry waste: earthworm growth, plant-available nutrient and microbial quality of end materials. Biores Technol 112:179–185

Srivastav RK, Gupta SK, Nigam KDP, Vasudevan P (1994) Treatment of chromium and nickel in wastewater by using aquatic plants. Water Res 28(7):1631–1638

Stirk WA, van Staden J (2003) Occurrence of cytokinin-like compounds in two aquatic ferns and their exudates. Environ Exp Bot 49(1):77–85

Swapna MM, Prakashkumar R, Anoop KP, Manju CN, Rajith NP (2011) A review on the medicinal and edible aspects of aquatic and wetland plants of India. J Med Plant Res 5(33):7163–7176

Tauseef SM, Tasneem A, Banupriya D, Vaishnavi G, Abbasi SA (2013) HEVSPAR: a novel vermireactor system for treating paper waste. Off J Pat Off 24:12726

Tauseef SM, Abbasi T, Banupriya G, Banupriya D, Abbasi SA (2014) A new machine for clean and rapid separation of vermicast, earthworms and undigested substrate in vermicomposting systems. J Environ Sci Eng 56(4):495–498

Tengfei Z, Chaohui W, Baogang M, Zhuangde J (2016) Biomimetics studies of *Salvinia molesta* for fabrication. Micro Nano Lett 11(6):291–294

Toerien DF, Cary PR, Finlayson CM, Mitchell DS, Weerts PGJ (1983) Growth models for Salvinia molesta. Aquat Bot 16(2):173–179

Verma DK, Hasan SH, Banik RM (2016) Photo-catalyzed and phyto-mediated rapid green synthesis of silver nanoparticles using herbal extract of *Salvina molesta* and its antimicrobial efficacy. J Photochem Photobiol B: Biol 155:51–59

Yang CY, Yang CY, Sung CK (2013) Enhancing air retention by biomimicking Salvinia molesta structures. Jpn J Appl Phys 52(6S):06GF08

Zamorano M, Pérez JIP, Pavés IA, Ridao ÁR (2007) Study of the energy potential of the biogas produced by an urban waste landfill in Southern Spain. Renew Sustain Energy Rev 11(5):909–922

Zeiger C, da Silva ICR, Mail M, Kavalenka MN, Barthlott W, Holscher H (2016) Microstructures of superhydrophobic plant leaves-inspiration for efficient oil spill cleanup materials. Bioinspiration Biomim 11(5):056003

Microplastic Contamination and Life Cycle Assessment of Bottled Drinking Water

Dharnidhar Choudhary, Caneon Kurien and Ajay Kumar Srivastava

Abstract Microplastic released from bottled water is found to be an anthropogenic contaminant affecting human health. Microplastic particles can lead to oxidative stress if it enters the human tissue. Most microplastic particles are found to be in size range of 5–10 μm and it can be detected effectively by using Micro-Raman Spectroscopy. One litre of bottled water is found to contain an average of 10,000 microplastic particles. Source for release of microplastic particles in bottled water is Polypropylene and Polyethylene terephthalate (PET) which is the major ingredient of bottle cap material. In this paper, a detailed review on the microplastic content of bottled water has been carried out and also an assessment on the life cycle of bottled drinking water is performed. The results of this study concluded that the harmful effects of microplastic particles in bottled drinking water can be reduced by using Polylactic acid (PLA) which has low impacts on carcinogenic and respiratory organics for bottle cap production.

Keywords Bottled water · Microplastic · Polyethylene terephthalate (PET) · Polypropylene · Polylactic acid (PLA) · Micro-Raman spectroscopy

1 Introduction

Microplastic contamination of water resources due to solid waste disposal has gained attention due to its impact on the environment and human health (Schweitzer and Noblet 2018). Recently it has been identified that bottled drinking water which is

D. Choudhary · C. Kurien (✉) · A. K. Srivastava
Mechanical Engineering Department, University of Petroleum and Energy Studies, Dehradun, India
e-mail: ckurien@ddn.upes.ac.in

D. Choudhary
e-mail: 500054211@stu.upes.ac.in

A. K. Srivastava
e-mail: akumar@ddn.upes.ac.in

N. A. Siddiqui et al. (eds.), *Advances in Water Pollution Monitoring and Control*, Springer Transactions in Civil and Environmental Engineering,
https://doi.org/10.1007/978-981-32-9956-6_4

claimed to be pure is contaminated by plastic particles released from the containers (Lahens et al. 2018). Plastic is defined as any polymer, which is composed of various elements such as carbon, hydrogen, nitrogen, chlorine and sulphur with thermoplastic properties. The plastic production increases exponentially since its entrance has accepted towards consumer stage (Crawford and Quinn 2017). Some features of plastic which attract the manufacturers and the buyers are its light-weight allowing it for easy transportation and resistance to breakage. With the rise in plastic production, there has been associated rise in plastic pollution to the external environment and also health-related issues due to the presence of tiny (micro) plastic particles inside bottled drinking water (Oßmann et al. 2018). Studies conducted by the World Health Organization (WHO) confirmed the presence of plastic particles in bottled drinking water and found traces of microplastic contaminants in all samples of commercially available packed drinking water. Microplastic particle lays a harmful effect on human health and is carcinogenic in nature. Polypropylene and Polyethylene terephthalate (PET) also lead to contamination of bottled water and make it unhealthy for drinking purpose. PET is a long-chain polymer of polyester family (Jerpdal et al. 2018).

Microplastic has the ability to pass through human body, although very tiny particles may be absorbed by organs, such as kidneys and liver which adversely affect its functioning and have wide impact on metabolism (Carbery et al. 2018). Since in developing countries like India, Mongolia and Pakistan there have been scarcity of drinking water from past decades, which gained the trust of public on these bottled drinking water. Also, bottled water remains a necessity to the remote areas with lack of access to drinking water. Along with the contamination of bottled water, the other area to be considered is the life cycle of a drinking water bottle. As we are aware of the time within which we are advised to use the bottled water, after which it is claimed unhealthy from the manufacturer point of view. Even after that these bottles lay a very adverse effect on environment as they are non-biodegradable. Reusing the bottles instead of manufacturing new ones is the profitable business for bottle manufacturers, which ultimately affects the bottle life along with increased number of microplastic in bottled water. Helmholtz centre for Polar and Marine Research conducted an investigation on the presence of microplastics in raw and tap water with the help of micro FTIR where the results showed the presence of particles in the size range of 50–150 μm (Mintenig et al. 2014). It was predicted from the results that the source of these particles would be the plastic pipe and fittings used for transportation of water. Major factors considered to have an influence in the microplastic contamination are production date, material used for bottle, bottle cap and also exposure to varying temperature conditions. In this paper, a detailed review on the presence of microplastic in bottled drinking water as well as its impact on human health and also an assessment on life cycle of bottled drinking water has been carried out.

2 Materials and Methods

2.1 Materials

One of the most effective techniques to identify the amount of microplastic in bottled water is micro-Fourier Transform Infrared Spectroscopy (μ-FTIR), which has the ability to detect particle size more than 20 μm. The Fourier-transform spectroscopy technique can be used for identification of the functional group of the particle present in sample. Advanced instrument for counting the number of particles present in the sample is Single Particle Explorer, which has the ability to count a maximum of 5000 particles present in the sample. Another technique used is the Micro-Raman Spectroscopy, in which it can detect even smaller particle sizes up to 1 μm. In this paper, a detailed study on the effectiveness of Micro-Raman Spectroscopy in detecting the amount of microparticle present in bottled water has been conducted. Micro-Raman Spectroscopy technique uses a spectrometer to measure the vibrational spectra of microscopic samples. Raman microscope is integrated with Raman Spectrometer. In this technique, laser is being imposed on sample by passing through filtered probe and then to lens, the reflected rays are then absorbed by spectrograph and then passed on to CCD detector to read variation in the wavelength of particles as shown in Fig. 1.

In a study carried out by (Schymanski et al. 2017), packed water bottles were collected from different manufacturers and tested to check the amount of microplastics. The study showed that an average of 10 plastic particles was present in 1 litre of sample and with average width, higher than that of human hair. The study also pointed out that these tiny particles of plastic gets into bottles from surrounding, while mounting the bottle cap from machines and while opening the bottles these

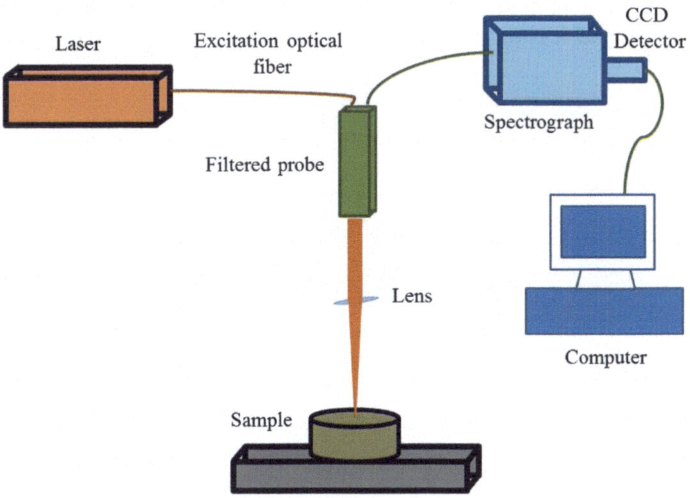

Fig. 1 Working principle of Raman Spectroscopy

particles get mixed with the water and make the water contaminated. Water samples also must be classified into categories based on the presence of carbonic acid in the sample.

2.2 Methods

Exciting lasers at different wavelengths of the microscopic sample can do the analysis for vibrational spectra of the microscopic samples, so that spectrometer can collect the presence of particle in it. Raman Spectroscopy works on the basis of inelastic scattering of light which occurs when a photon on microscopic sample molecule excites from its vibrational state to higher virtual state and then comes down to lower energy level which is at a slightly higher than the initial energy level, this shift in the wave number is known as the Raman shift.

A laser of monochromatic light is generated and is flashed on sample, passing through a Laser Transmitting Filter as shown in Fig. 2. Some of the light gets absorbed by the filtre while the remaining is passed through it and strikes on sample. Here the light is dispersed in different directions, so another filtre is used here for blocking unwanted reflected rays and the light rays which are passed through this filtre are collected by spectrometer for analysis of microparticles and finally are reflected as an output to the attached computer. The major functional group of microplastics observed in bottled water belongs to the functional group of polyethylene terephthalate and polyester. The percentages of different plastic components present in the water sample from returnable and one-way bottles are shown in Fig. 3. While performing Raman Spectroscopy it was observed that when monochromatic light (laser) is passed through the microscopic sample, some of the light is absorbed by the sample and the microplastic particles due to inelastic collision while the remaining light is reflected back and it is then passed through filtre, here unwanted light particle is

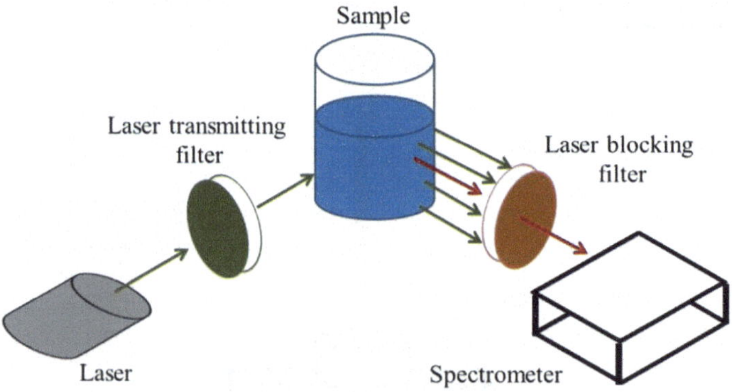

Fig. 2 Set-up for Raman Spectroscopy

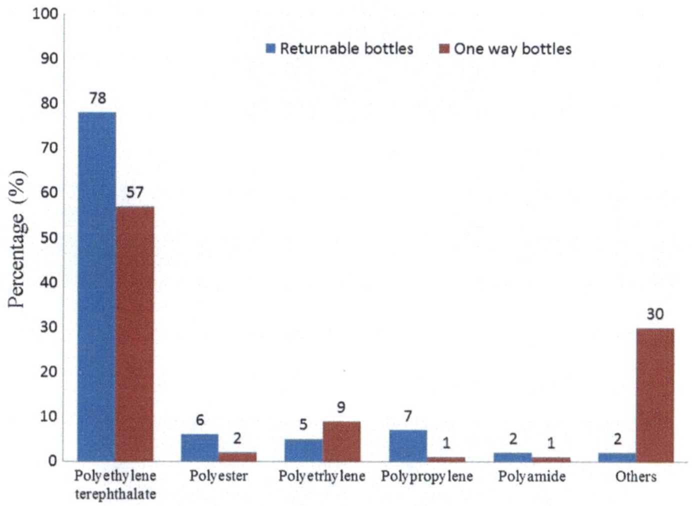

Fig. 3 Percentage of plastic components in one way and returnable bottles

differentiated and remaining are sent to (Charged Coupled Device) CCD and then to computer to obtain graph.

3 Life Cycle Assessment of Bottled Drinking Water

Environmental concerns have been rising since plastic products came into existence, due to its non-biodegradable property (Horowitz et al. 2018). Even though plastic is non-biodegradable but it can be recycled for which energy is used to sort and process the plastic which ultimately increases the cost (Bui et al. 2018). As recycling of bottles increases the manufacturing cost, so industries use new plastic resin, which is eventually harmful to the environment. There are certain type of bottles like PLA, recycled PET and regular PET, these are used for the purpose of water packing (Chapa-Martínez et al. 2016). Keeping these factors in mind, corn-based Polylactic acid (PLA) may be used to manufacture bottles because PLA is derived from renewable resources and it consumes energy, which directly increases the life cycle of a PLA bottle (Lucchetti et al. 2016). Various stages involved in the life cycle of recyclable plastic bottles are as shown in Fig. 4.

Life cycle of a bottle depends on certain factors, which includes transportation, temperature subjected and pressure applied to bottles. Transportation of bottle plays a very major role as the final condition of bottle majorly depends on how the bottle is transported from the manufacturing unit to the consumers (Chen et al. 2016). If the bottles are transported in a well-arranged manner then the mentioned life period of a bottled in maintained otherwise lifespan of bottle is decreased. Second,

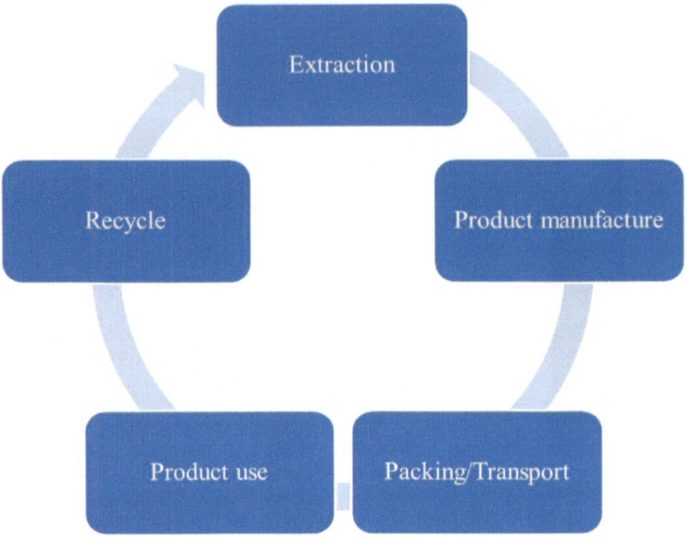

Fig. 4 Life cycle of recyclable plastic bottles

if the bottle is subjected to high temperature for regular time, then there may be deformation in the shape of bottle which eventually deteriorates its quality and also decreases bottle lifetime span (Garfí et al. 2016). Third, if bottle is subjected to continuous pressure then there may be change of slight reduction in life period of bottle as due to pressurizing the bottle it deforms and does not fulfil its purpose (Yan et al. 2018). Micro-sized plastic particles were found and even some of them are of the size of human hair. Microplastic such as Polypropylene (PE), Polystyrene (PS), PET and nylon particles are also detected to be present in bottled water (Tao et al. 2018). PET is also found to absorb flavour and odour from material inside the bottles. It is commonly recyclable. The Polypropylene is the material which is used to manufacture the bottle caps (Axel Plastics present new mould release for bottle caps, 2017). The main reason for the contamination of bottled water is (Polypropylene) observed as the microplastic particles gets into bottle while opening and closing it and while doing packing very tiny plastic particle present in surrounding gets into the bottle which ultimately contaminate the water. The manufacturing stage, distribution stage and storage stage are the major stages which have a direct impact on the life of packed bottles (Simon et al. 2016). The operating system during these stages has to be made more sustainable to reduce the contamination of bottled water and also to reduce the impact on environment.

4 Conclusions

Microplastic contamination is found to occur in bottled drinking water and around 70% of particle sized between 5 and 20 microns. Consumption of bottled water would lead to severe health-related problems. The presence of these particles can be analyzed by particle analyzers and by conducting spectroscopy. The study analyzes that recycled bottle contains more number of microplastic particle in its packing as compared to the first time manufacturing of bottles containing plastic. The microplastic contamination has been observed to occur in plastic returnable bottles, one-way bottles and glass bottles also. The size of these particles is in the range of 1–20 μm and the number of particles is in the range of 10–180 particles per litre. Microplastic constituents can be lowered in bottled water by using Polylactic acid (PLA) for manufacturing bottle cap as mainly cap of bottles lead to contamination water. Also, health-related issues can be controlled to some extend by using glass bottles for drinking purpose instead of plastic bottles. The future scope of this work includes the analysis of the microplastic contaminants in bottled drinking water by using particle size analyzer and the functional groups of the contaminants by conducting Fourier-transform infrared spectroscopy.

References

Axel Plastics presents new mould release for bottle caps. Addit Polym 2017(11):3–4 (2017)

Bui NK, Satomi T, Takahashi H (2018) Recycling woven plastic sack waste and PET bottle waste as fiber in recycled aggregate concrete: an experimental study. Waste Manag 78:79–93

Carbery M, O'Connor W, Palanisami T (2018) Trophic transfer of microplastics and mixed contaminants in the marine food web and implications for human health. Environ Int 115:400–409

Chapa-Martínez CA, Hinojosa-Reyes L, Hernández-Ramírez A, Ruiz-Ruiz E, Maya-Treviño L, Guzmán-Mar JL (2016) An evaluation of the migration of antimony from polyethylene terephthalate (PET) plastic used for bottled drinking water. Sci Total Environ 565:511–518

Chen L, Pelton REO, Smith TM (2016) Comparative life cycle assessment of fossil and bio-based polyethylene terephthalate (PET) bottles. J Clean Prod 137:667–676

Crawford CB, Quinn B (2017) Plastic production, waste and legislation. In: Crawford CB, Quinn B (eds) Microplastic pollutants. Elsevier, pp 39–56

Garfí M, Cadena E, Sanchez-Ramos D, Ferrer I (2016) Life cycle assessment of drinking water: comparing conventional water treatment, reverse osmosis and mineral water in glass and plastic bottles. J Clean Prod 137:997–1003

Horowitz N, Frago J, Mu D (2018) Life cycle assessment of bottled water : a case study of Green2O products. Waste Manag

Jerpdal L, Åkermo M, Ståhlberg D, Herzig A (2018) Process induced shape distortions of self-reinforced poly(ethylene terephthalate) composites. Compos Struct 193:29–34

Lahens L et al (2018) Macroplastic and microplastic contamination assessment of a tropical river (Saigon River, Vietnam) transversed by a developing megacity. Environ Pollut 236:661–671

Lucchetti C, De Simone G, Galli G, Tuccimei P (2016) Evaluating radon loss from water during storage in standard PET, bio-based PET, and PLA bottles. Radiat Meas 84:1–8

Mintenig S, Löder M, Gerdts G (2014) Mikroplastik in Trinkwasser, Untersuchung im Trinkwasserversorgungsgebiet des Oldenburgisch-Ostfriesischen Wasserverbandes (OOWV) in Niedersachsen, Probenanalyse mittels Mikro-FTIR Spektroskopie. Helgol. Oldenburgisch-Ostfriesischer Wasserverband (OOWV). Alfred-Wegener-Institut

Oßmann BE, Sarau G, Holtmannspötter H, Pischetsrieder M, Christiansen SH, Dicke W (2018) Small-sized microplastics and pigmented particles in bottled mineral water. Water Res

Schweitzer L, Noblet J (2018) Water contamination and pollution (Chapter 3.6). In: Török B, Dransfield T (eds) Green chemistry. Elsevier, pp 261–290

Schymanski D, Goldbeck C, Humpf H, Fürst P (2017) Analysis of microplastics in water by micro-Raman spectroscopy: release of plastic particles from different packaging into mineral water. Water Res

Simon B, Ben Amor M, Földényi R (2016) Life cycle impact assessment of beverage packaging systems: focus on the collection of post-consumer bottles. J Clean Prod 112:238–248

Tao Y et al (2018) Thermodynamic sorption properties, water plasticizing effect and particle characteristics of blueberry powders produced from juices, fruits and pomaces. Powder Technol 323:208–218

Yan X, Ward S, Butler D, Daly B (2018) Performance assessment and life cycle analysis of potable water production from harvested rainwater by a decentralized system. J Clean Prod 172:2167–2173

Impact of Water Pollution on Endemic Species

Ambika Yadav, N. A. Siddiqui and Abhishek Nandan

Abstract The review paper aims at the 'Impact of Water Pollution on Endemic Species'. This paper is an attempt to explain the reasons due to which species are getting endemic as a result of changes caused in their organs. The main cause is anthropogenic activities. This paper tells us how the human activities and exposure of toxicants causing harm to these kinds of species directly and indirectly results into negative consequences for the aquatic organisms through changes in their cells towards the population of the species. The endemic species 'Dawkinsia Tambraparniei' are found near the areas of River Tamiraparani (Dahanukar 2015). These types of species are under a great threat due to man-made activities which are increasing day by day. To identify this type of crisis of the species, it is observed by testing from the five different pollutant locations of the river. There are examinations that are taken out from the internal organs such as the brain, gill, heart, kidney and liver. Most of the organs of the fish damages are highly observed in Kokkirakulam and Vannarapettai areas. The species have same genus but they have different subspecies also. The important measures are taken to save the endemic species. Continuous effective steps in inhabited zones are also taken.

Keywords Endemic species · Genes · Pollutants and tissues

1 Introduction

Water pollution is the contamination of water bodies. Pollutants are directly or indirectly discharged into water bodies without adequate treatment to remove harmful compounds. Sewage and garbage and liquid waste of households and factories are discharged into lakes and rivers which make water poisonous for aquatic organisms. Industrial waste is the waste produced by industrial activity which includes any material that is rendered useless during a manufacturing process such as that of factories, industries, mills and mining operations. Endemic species are the organ-

---

A. Yadav · N. A. Siddiqui · A. Nandan (✉)
University of Petroleum and Energy Studies, Dehradun, India
e-mail: abhisheknandan24@gmail.com

N. A. Siddiqui et al. (eds.), *Advances in Water Pollution Monitoring and Control*,
Springer Transactions in Civil and Environmental Engineering,
https://doi.org/10.1007/978-981-32-9956-6_6

isms that exist only in one geographic region. Here the review paper is about the five different locations with species getting endemic due to land use patterns such as discharge of sewage, industrial effluents and paper mill discharging, etc. The steps have been taken to conserve our natural ecosystem from anthropogenic pollutants such as heavy metals into the freshwater (Mohammed and Al-kahtani 2009). The heavy metals from automobiles, restricted waste and industrial wastes which are harmful to our environment as they cause changes in the aquatic systems where the organisms survive (Abdullah et al. 2008). A large amount of collection of heavy metals in aquatic organisms gives us an idea about the health of organisms. There are certain changes which have been noticed in the body of the organisms and the reason could be that contamination which slowly and steadily degrades organism's body structure. As of now, there are certain drastic changes in the different parts of the fish which could give a better idea as those changes are noticeable and this kind of changes could be easily measured and hence the appropriate steps can be taken. Now, the different types of changes can be noticed and determined in different situations through the techniques which were used in ancient times. There are some organs such as gills, liver and kidney which have the ability to gather harmful substances. These organs absorb substances at higher levels. Now following functions of organs are described. The heart of the fish plays a vital role in the blood circulation (oxygenated blood is transported to different parts of the body), transportation of the essential nutrients as well as oxygen. The function of the gills is that it allows the fish to get oxygen as well as it acts as the lungs to them. The liver plays a very important role in breaking down fats with the help of enzymes. Apart from this, the liver also helps in destroying the old blood cells. As the amount of dangerous substances increase in the body of the fish, there is a strong destruction in their structure which is a major concern for the environment. So finally, this information is quite useful to tell about the changes which occur in the body of fish by observing the different levels of changes in the species of fishes after taking certain records for a longer period of time interval. This also gives a better idea about the changes taking place in other species at an alarming level. Different patterns of the genes give information about the wide variety of fishes and other species related to them with different and unique characteristics of their body. A river named as 'Tamiraparani' consisting of 'Dawkinsia Tambraparniei', which is a type of endangered fish that has been used to find out the extent of water pollution on them and how they are affected from it, causing highly adverse conditions for the other organisms and ecosystem.

2 Review of Literature

Due to the amount of different and impure substances present in the polluted water, there are certain changes which occur in different forms and are found in the genes of species which results into change in the molecular level of DNA (Frenzilli et al. 2004; Walker et al. 2006). Dawkinsia Tambraparniei has been chosen as a species which is used for observing the changes in the pattern of genes at those selected different five sites which are taken into account. Hebert et al. (2004) and Pereira et al. (2013) observed that the changes in the members of the different species are far better than the changes observed in the members of the same species which helps in differentiating species with the help of a sequence of a compounds consisting of nucleoside group combined with the phosphate group in the chain of mitochondrial cytochrome. This results in the making of DNA (Isenbarger et al. 2008). According to Ward (2009), samples of D. Tambraparniei were taken and as a result, it was observed that difference between the type of species taken at different five locations differ by only approximately two base pairs which clearly states the similarities between them and tells that the all the five species are similar to each other with respect to their characteristics. The nucleotides act as the most important key to identify the changes observed in the genes because they act as an electron carrier and carries metabolic energy. Apart from this, it activates the messaging signals in the body which makes the cell activities active and lead them to work (Lee 2000) and emerging of different kinds of suitable relationships in classifying the genes arrangement in different species (Brown et al. 1979). It has been found that certain number of changes between two molecules have decreased the process of gradual change and development in the genes over successive generations (Isenbarger et al. 2008). The difference in the genes was accepted because the population of these kinds of species has been kept apart due to the man-made harmful activities in the bottom of the stream (Harrison et al. 2012).

3 Methodology Adopted

3.1 Different Locations for the Collection of Sample

Peninsular India has one of the famous rivers named as 'Tamiraparani'. In Southern Western Ghats, this river flows from certain areas of Tamil Nadu. The area of land drained by this river consists of moderate to poor soils. The area of land drained by this river consists of loose soil, very fine topsoil and rough soil which could be easily noticeable observed during the days of Summer season which probably shows the erosion impacts on the riverbed. The river has been polluted to a greater extent slowly and slowly due to the dumping of domestic as well as industrial wastage. One of the major reasons for this river to be polluted is the accumulation of the waste coming from the industries that could be from chemical industries as well as manufacturing industries. So this has led to the increase in water pollution which ultimately leads to the death of living organisms or genetic changes in aquatic organisms (Table 1) (Dinakaran and Anbalagan 2007; Johnson and Arunachalam 2009; Magesh and Chandrasekar 2014; Kaliraj et al. 2015).

Table 1 Sampling at the river location

Sampling locations	Number of species of single type	Height (m)	Purpose of land
Sample 1	5	80	Wastage from homes, farming and industrial wastage
Sample 2	7	63	Rural and urban area, wastage from homes and farming
Sample 3	5	52	Rural and urban area and industrial wastage
Sample 4	6	59	Rural and urban area, wastage coming from small and large scale industries
Sample 5	5	36	Reservoirs and farming

3.2 *Study of Species Dawkinsia Tambraparniei*

It is an endangered species which lives in the Tambraparniei. It is declining due to activities by human as well as confined regions. It is categorized as one the endangered species. It is a fish which is quite small in size and it flows moderately and consists of bedrocks and sandy bottom. Length of the body is 9.2 cm. It has fine body shape and attractive fin colour. This species is found in tributaries, from the middle of the river where the water flows fastest, from where the water stretches at a lower level over an area and related by lands. The range within which this type of species is found is often less than 97 km approximately.

3.3 *Fish Sampling*

The samples of this fish species are collected during dry season and that also one time from different habitats of five sites. These are taken with the help of net that has single filtration with four square openings across one inch of screen of approximately 7–11 mm. For the molecular analysis, small size of the body of dorsal fish was kept in ethanol.

3.4 *Study of Structures Under the Microscope*

The vital organs of the species are taken into account for testing. The samples are fixed in 10% neutral buffer formalin for 48 h as well as samples and moisture are removed from them by using 2-propanol in a different concentration of ethanol series one by one for 1 hour. In two jars, there were xylene which was kept for almost 1 hour. After that it was then put into wax named as paraffin with the help of L-mould. Organs are sectioned at 4–5 μm thickness. Tissues are put on glass slides under microscope and albumin solution was put into it. These slides are then dried overnight. When they are dried, haematoxylin and eosin solutions are put over it. These are observed under microscope for further observations.

3.5 *Extracting DNA*

The chromosomal DNA is separated from a device which is used to don the models without any bend down by using a tissue kit named as NucleoSpin. Tissues are then placed in a 1.5 ml tube. 180 ml of Titanium buffer and 25 μl of enzyme known as proteinase K or 1 M Dithiothreitol is added and heated at 52 °C in a hot water placed in an oven until the membrane of the cell is completely broken down. After this, large group of hydrolytic enzyme of approximately 4 μl is mixed and heated at 25 °C for 4 min. 210 μl of 100% ethanol are added and mixed thoroughly by vortexing (a whirling mass of solution formed). After that, the mixture was pipette into NucleoSpin tissue column placed in a 2 ml tube and centrifuged (rotating at high speed) for 1 min. The tissue column is transferred to a new 2 ml tube. It is washed with 500 μl of concentrated wash buffer solution. Now again it is washed using 600 μl of different buffer solution such as elution buffer. After washing, tissue column is placed in a clean 1.5 ml tube and DNA is eluted (to remove one material from another using a solvent) using 50 μl of elution buffer solution. The obtained DNA sample is stored at −19 °C.

3.6 *Increasing the Number of Copies of Gene (Amplification)*

The cytochrome oxidase subunit 1 gene is increasing in the number of copies of a gene by PCR (polymerase chain reactions) carried out in a 20 μl reaction volume which contain different buffer solutions, enzymes and primers. The reactions are further completed in a cycler that is used to amplify the segments of DNA using analyzers.

3.7 *Sequence Analysis*

The phylogenetic analysis of COI gene sequence data is performed using MEGA5 software package. Final work of sequences is done with the help of a machine which edits the sequences in a far good way and arranged structure.

4 Result and Discussion

4.1 Testing of Brain Tissue

At sample site 1, the brain sample consists of various changes such as clumsiness in fibre named as Purkinje. At sample site 2, the brain sample consists of various changes like damaged and small broken pieces of nerves cells as well as damage to the nervous system. At sample site 3, notable changes such as damage to the central nervous system are observed in the brain. At sample site 4 and 5, the brain sample consists of various changes such as an abnormal clustering of gill cells around neuronal cells. Through this information related to brain was observed such as death of cells or tissues through injury, brain haemorrhage and inflammation of the membranes surrounding brain and spinal cord as a result of lead effects which was too dangerous. Lead acts to be a blood–brain barrier which causes accumulation of fluid in the intracellular and extracellular spaces of brain which is really a fatal condition. Some of the adverse changes observed in the brain of the fish that consumes heavy metal are swelling of cells due to injury, cellular damage in the interior of the brain and lobes of the brain, non-functioning of nerve cell, stiffening and death of cells in the brain.

4.2 Testing of Gill Tissue

Gills are those parts of the fish which come in direct contact with the pollutants. The sample of the gill taken from site location 1 shows various changes such as goblet which secrete mucous, degenerative alteration and areas which do not have any structures. At location 2, variations are observed in the cartridge of fish which shows abnormal structure, congested blood vessels of clumpsy adhesion and blood congestion in the gills. At location 3, vasodilatation (stiffening) of blood vessels in secondary lamella, edema (swelling caused by excessive fluid in the tissues of gill) and lamellar aneurysm (accumulation of blood vessels in the secondary lamellae) are observed. In location 4, shrinking of gill tissues of lamella with slightly observed curls at the end and necrotic cell death (death of cells) are observed. At location 5, fusion of secondary lamella (patristic infection on gills), death of cells and accumulation of inflammatory cells are observed (Borges et al. 2003).

4.3 Testing of Heart Tissue

There is a structure of cardiac muscle fibres with large nuclei and normally arranged permanent cells for the function of cell in the heart. At location 1, fragmentation (process of breaking or being broken) of myocardial fibres is observed. Similar fragmentations are observed in rest of the other sites. At location 5, hypertrophic changes (increase in the volume of an organ due to the enlargement of its component cells) in blood vessels are observed (Ahmad et al. 2011).

4.4 Testing of Kidney Tissue

The nephron cells are usually found in the kidney of fish. A pair of renal corpuscle and renal tubule is generally found in the cell. Homeostasis maintenance is necessary in the kidney for performing various functions such as the wastes removal, reabsorption to maintain the pH, to increase the volume of blood fluids and erythropoiesis (process which produces red blood cells). At location 1, cloudy swelling of hematopoietic (formation of blood cellular components) tissues is noted. At location 2, changes such as irreversible condensation of chromatin in the nucleus of a cell, disorganized tubules and necrotic damages are observed. At location 3, portion of kidney nephron between collecting tubule and hyaline degeneration of tubular epithelium (formation of rounded masses of acidophilic substances that have a glassy appearance) are observed. At location 4, melano-macrophage (pigmented phagocytes found primarily in the kidney) is collected between the renal tubules and cellular degeneration in tubules. At location 5, nuclear hypertrophy (increased biosynthetic activities in myocytes of kidney from the large amount of exposure of cadmium) and distal tubules degeneration are noted.

4.5 Testing of Liver Tissue

The arrangement of polyhedral symmetry of hepatocytes (liver cell) around the central vein was shown in the normal type of liver. Hepatocytes consist of the appearance of cells that are stained by pink dye. In the open-pore capillary, there was a distribution of the blood cells. At location 1, liver tissues observed the enlargement of pancreatic tissue, leucocytes collection and dilation of sinusoids (found in the absence of impaired blood flow). At location 2, bile stagnation (a condition where bile cannot flow from liver), haemorrhage and melano-macrophage aggregate are identified. At location 3, dilated blood vessels and changes due to enzymatic inhibition and damage to the cellular membrane are recognized (Groopman et al. 1985). At location 4, cytoplasmic degeneration and necrotic changes observed the poor conditions of river. At location 5, haemorrhage, patchy degeneration (old degeneration of the fibrils), blood congestion (a condition where veins swell up and blood collects) and distended sinusoids are identified. The pollutants are collected in the fatty tissues of fish such as liver and the consequences become quite adverse when the pollutant concentrations pass through a certain level in the liver. Hepatocytes vacuolization, hepatic cirrhosis (diseased liver damage), necrosis (cell injury), shrinking, degeneration of parenchyma, nuclear death of cells and increase of vacant spaces are the certain changes observed in the liver of the fish which contains lead in a very large amount.

5 Conclusion

From the review paper, I would like to conclude that due to adverse toxicant effect on the body of species, there is a change in the cellular level as well as certain drastic changes could be observed in the internal organs of the fish. Due to the anthropogenic activities like throwing of domestic sewages, industrial waste, waste coming from agriculture, mining activities, dumping of waste, oil leakage from chemical factories and burning of fossil fuels there is presence of heavy metals in the water which is consumed by the fishes and aquatic organisms. After examining tissues of the fish, results are quite negative as heavy metals lead to chronic contaminations with adverse effect for the aquatic organisms through changes in their genetic level as well as cellular level. In addition to this, after doing a majority of tests on different parts of the body of fish, a conclusion came that all the fishes belongs to a same group but they have variations in their genetic level. The toxicants enter into their body and due to their long-time exposure in the water, slowly and slowly their body starts consuming those heavy metal which in a way starts degrading their body and this leads to their extinction from the environment.

References

Abdullah E-G, Mehanna EE, Meki A-R (2008) Evaluation of lead and cadmium levels in fresh water fish farms at Qassim region, K.S.A. J Agric Vet

Ahmad B, Qureshi TA, Manohar S, Kaur P, Khaliq R (2011) Effect of cadmium chloride on the histoarchitecture of liver and kidney of a freshwater catfish, Clarias batrachus. Int J Environ

Borges G, Mendonca P, Joaquim N, Coucelo J, Aureliano M (2003) Acute effects of vanadate oligomers on heart, kidney, and liver histology in the Lusitanian toadfish (Halobatrachus didactylus). Arch Environ Contam Toxicol

Brown WM, George M Jr, Wilson AC (1979) Rapid evolution of animal mitochondrial DNA. Proc Natl Acad

Dahanukar N (2015) Dawkinsia Tambraparniei. The IUCN Red List of Threatened Species 2015

Dinakaran S, Anbalagan S (2007) Anthropogenic impacts on aquatic insects in six streams of South Western Ghats. J Insect Sci 7(37):1–9

Frenzilli G, Scarcelli V, Del Barga I, Nigro M, Bolognesi FL, Sturve J (2004) DNA damage in eelpout (Zoarces viviparus) from Goteborg harbour

Groopman JD, Wolff T, Distlerath LM (1985) Substrate specificity of human liver cytochrome p-450 debrisoquine 4-hydroxylase probed using immunochemical inhibition and chemical modeling

Harrison TM, Grove M, Lovera OM, Catlos EJ (2012) A model for the origin of Himalayan anatexis and inverted metamorphism. J Geohys Res

Hebert PDN, Penton EH, Burns JM, Janzen DH, and Hallwachs W (2004) ten species in one: DNA barcoding reveals cryptic species in the Neotropical skipper butterfly Astraptes fulgerator. Proc Natl Acad

Isenbarger TA, Finney M, Velazquez CR (2008) Miniprimer PCR, a new lens for viewing the microbial world. Appl Environ Microbiol

Johnson J, Arunachalam M (2009) Diversity, distribution and assemblage structure of fishes in streams of Southern Western Ghats, India. J Threatened Taxa 507–513

Kaliraj S, Chandrasekar N, Magesh NS (2015) Morphometric analysis of the River Thamirabarani sub-basin in Kanyakumari District, South west coast of Tamil Nadu, India, using remote sensing and GIS. Environ Earth Sci 73(11):7375–7401

Lee YH (2000) Evolution of sea urchin Strongylocentrotus intermedius based on DNA sequences of a mitochondrial gene, cytochrome c oxidase subunit I. J Kor Soc Ocean

Magesh NS, Chandrasekar N (2014) GIS model-based morphometric evaluation of Tamiraparani subbasin, Tirunelveli district, Tamil Nadu, India. Arab J Geosci 7(1):131–141

Mohammed A, Al-kahtani (2009) Accumulation of heavy metals in tilapia fish Oreochromis niloticus from Al-khadoud spring, Al-hassan, Saudi Arabia

Pereira LHG, Hanner R, Foresti F, Oliveira C (2013) Can DNA barcoding accurately discriminate mega diverse Neotropical freshwater fish fauna

Walker CH, Hopkin SP, Sibly RM, Peakall DB (2006) Principles of ecotoxicology, 3rd edn. Taylor & Francis, London; New York

Ward RD (2009) DNA barcode divergence among species and genera of birds and fishes. Mol Ecol Resour

Physiological Features of Rice Cultivars Which Influence the Quantity of Methane Emitted by the Cultivars

C. Luithui, Tabassum-Abbasi, Tasneem Abbasi and S. A. Abbasi

Abstract Intense current attention is focused on identifying factors which influence the extent of the emissions of greenhouse gases (GHGs) from paddy cultivation. Whereas it is now well established that the quantity of organic matter present in the soil coupled with favourable soil moisture and anoxic conditions are the principal drivers of methane emissions from paddy fields, the role of rice cultivars is much less documented or understood. Considering that even though modifications in the agricultural practices, including water management, can reduce methane generation in paddy fields, the effect of such initiatives can be reduced if the cultivar is such that it quickly and maximally transports methane to the atmosphere. Hence the physiology of the cultivar which influences the cultivar's tendency to emit methane is of great importance. This paper summarizes the available knowledge on this subject.

Keywords Rice · Paddy · Cultivars · Methane · Global warming · Plant physiology

1 Introduction

The food grain of which largest quantities are generated and consumed in Asia is rice. An estimated 190 million hectares of land has been committed to rice cultivation across the world (FAOSTAT 2009). With global population rising rapidly, the need for producing larger quantities of rice will increase in proportion (Abbasi et al. 2019; Dhanuja et al. 2020). Simultaneously, the world is confronted with the

C. Luithui · T. Abbasi · S. A. Abbasi (✉)
Centre for Pollution Control and Environmental Engineering, Pondicherry University, Puducherry 605014, India
e-mail: abbbasi.cpee@gmail.com

T. Abbasi
e-mail: tasneem.abbasi@gmail.com

Tabassum-Abbasi
Department of Health, Safety, Environment and Civil Engineering, University of Petroleum and Energy Studies, Dehradun 248007, India

N. A. Siddiqui et al. (eds.), *Advances in Water Pollution Monitoring and Control*, Springer Transactions in Civil and Environmental Engineering,
https://doi.org/10.1007/978-981-32-9956-6_7

problem of global warming and rice production has been one of the major contributors to anthropogenic greenhouse gas emissions (GHGs). In terms of methane alone, paddy fields are estimated to contribute substantially anthropogenic radiative forcing (Abbasi et al. 2019; Dhanuja, et al. 2020). It is likely that there will also be a substantial contribution from another GHG—nitrous oxide (N_2O)—but studies on N_2O emissions are much fewer and the understanding of the factors which influence N_2O emissions is much less rigorous (Luithui et al. 2017; Tabassum-Abbasi 2020a, b).

As for methane, it is now well-understood that flooded soils which are rich in organic matter, of the type which are common in traditional rice cultivation, are conducive towards creating anoxic conditions which favour anaerobic digestion of organic matter (Abbasi et al. 2009, 2015). That, in turn, leads to the emission of 'biogas' of which roughly one-third is carbon dioxide and two-thirds is methane (Ganesh et al. 2005; Abbasi and Abbasi 2010, 2011). If anoxic conditions are not developed, the organic matter will decompose aerobically, generating the GHG carbon dioxide (CO_2). But the formation of methane (CH_4) is a much more harmful consequence vis–a-vis global warming than the formation of CO_2 because each molecule of CH_4 has 34 and 86 times greater global warming potential, on 100-year adn 10-year time horizons, respectively, than (IPCC 2013). But, even as a large number of studies have been done on the role of soil organic carbon and extent of flooding on methane emission of rice paddies, it is much less known about the role of rice cultivars on this phenomena. The present review aims to summarize and systematize the studies available in this area.

2 Methodology

An extensive search of the prior art with multiple keywords was made at the *Science Direct*, *Web of Science* and *Google* repositories for material on the subject of the review. As it threw only a few reports, further probing was done by seeing cross-references in various reviews earlier published on GHG emissions from paddy fields and by searching with more keywords. All the available reports have then been summarized and systematized, drawing conclusions.

3 Gist of the Findings

The available reports indicate that plant physiology plays an important role in CH_4 emission. All the components of the cultivar, viz. leaves, stem and root seem to be working in sync to augment each other's contribution to CH_4 emission (Fig. 1).

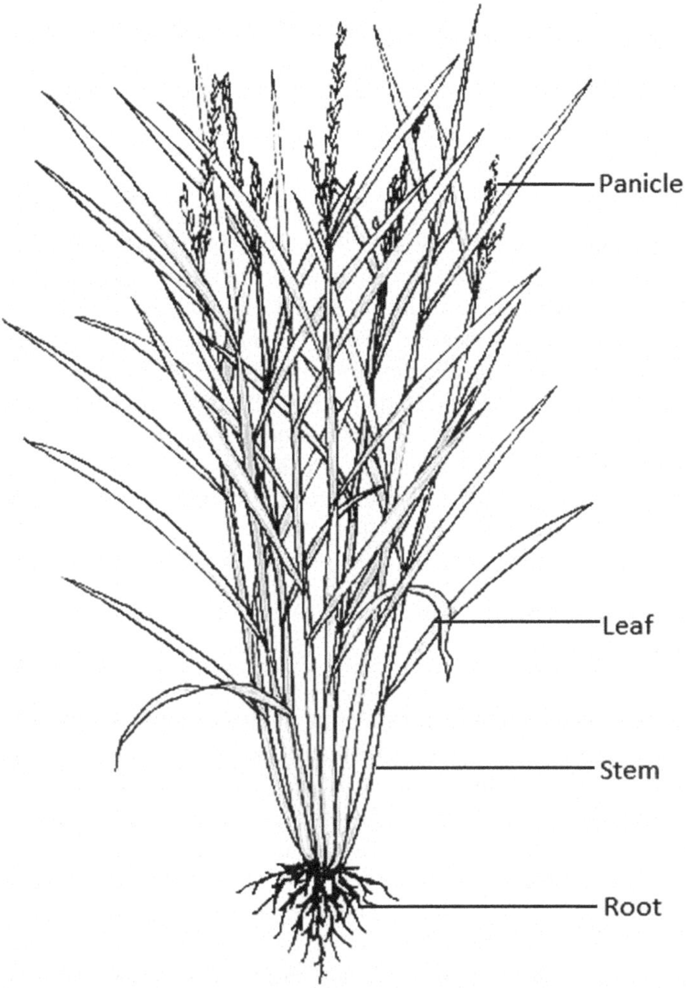

Fig. 1 A typical rice plant and its components which play a role in regulating methane emission

3.1 Role of Leaves

The leaf area and the number of leaves of a rice cultivar have been found to be positively correlated with CH_4 and N_2O emissions by Baruah et al. (2010). Indications have been found that stomata may be a significant site for releasing methane from the leaves (Das and Baruah 2008b). The stomata of the leaf sheaths have also been found to be a significant site for releasing CH_4 (Butterbach-Bahl et al. 2009). Since transpiration from the leaves occurs through the stomata, a cultivar with denser stomatal apertures in the leaves would result in higher CH_4 emissions. This also means that larger leaf area as well as a number of leaves can accommodate more number of stomata (Baruah et al. 2002; Gogoi et al. 2003), thereby enhancing a cultivar's tendency to emit CH_4.

3.2 Role of the Culm/Stem

The size of a cultivar's stem may directly influence methane emissions. Larger stem size can accommodate more amount of aerenchymatous tissues allowing more methane gas to be transported, resulting in higher CH_4 emissions (Ferch and Romheld 2001; Aulakh et al. 2002; Baruah et al. 2010). Larger stem size leads to larger vascular bundles, allowing a particular cultivar to emit more CH_4 by providing a conduit for the passage of the gas (Das and Baruah 2008a; Baruah et al. 2010). Larger xylem vessels also directly correlate with higher cumulative methane emissions. In addition, tiller number plays a role; a cultivar which has more tillers will cause higher emission of CH_4 and N_2O since tiller numbers are found to be positively correlated with CH_4 and N_2O emissions (Baruah et al. 2010).

3.3 Role of the Root

Size of the root is seen to be directly affecting methane emissions. During the early stages of crop growth, the conduit for methane transport from the soil to the atmosphere is limited because of the small size of the roots. As the root increases in size, the diffusion of CH_4 from the soil to the roots becomes more favourable resulting in higher CH_4 emissions (Baruah et al. 2010). Roots also provide a carbon source for the methanogenic bacteria in the form of root exudates; thereby supporting methane formation (Aulakh et al. 2000, 2001; Kimura et al. 2004; Baruah et al. 2010). As a rice plant approaches maturity, its roots begin to decompose and other plant tissues also suffer ageing. As a consequence, CH_4 emission declines very significantly in the maturity stage (Gogoi et al. 2008).

3.4 Role of the Overall Plant Biomass

Methane emission has been found to be positively correlated with plant biomass (Aulakh et al. 2002; Kerdchoechuen 2005; Gogoi et al. 2008; Baruah et al. 2010; Bharali et al. 2017a, b). Plant biomass increases 5–7 times from seedling stage to panicle initiation and flowering (Aulakh et al. 2002; Gogoi et al. 2008). This plays a significant role in increasing the ability of the plant to emit CH_4 from seedling stage till flowering (Gogoi et al. 2008). The general observation has been that larger the root length and the above-ground biomass, greater is the size of the leaves and stem, and in turn, higher CH_4 emission.

The xylem present in the vascular bundles of a cultivar may act as a passage for the CH_4 transport because a positive relationship between CH_4 emission and the rate of transpiration has been observed by Bharali et al. (2017a, b). In turn, transpiration plays a significant role in the rate at which CH_4 is emitted (Das and Baruah 2008a; Nisbet et al. 2009; Pangala et al. 2014; Bharali et al. 2017a, b). The transpiration rate is controlled by the number of stomata present in the leaf and size of the xylem vessel. As mentioned earlier, larger leaf size can accommodate more number of stomata. Also, it is apparent that rice cultivar with larger stem diameter can have larger xylem vessel. In short, greater the plant biomass, greater the number of stomata and the xylem size, boosting the CH_4 emissions.

3.5 Contribution from the Rate of Photosynthesis

Methane emission and the rate of photosynthesis appear to be positively correlated (Baruah et al. 2010). The latter increases with the increase in the density of the stomata since carbon is assimilated better by the plant (Bharali et al. 2017a, b; Leakey et al. 2009; Tanaka et al. 2013, 2014).

But this may not always be the case as the rate of photosynthesis alone does not determine methane flux, rather it is the totality of the processes associated with photosynthesis which does so.

An important factor is the allocation of the assimilated carbon to different parts of the cultivar. A cultivar with higher photosynthates allocated to the panicle would lead to better grain yield and lower methane emission than the cultivar that had translocated more number of photosynthates to the vegetative parts and the rhizosphere (Das and Baruah 2008b). In the latter case, the photosynthates from the vegetative parts and rhizosphere reach the soil and provide a substrate for methanogenesis (Jimenez and Lal 2006).

It has been estimated that about 11% of the assimilated carbon not translocated to the panicle region is emitted in the form of methane (Denier van der Gon et al. 2002).

4 Summary

As assessment of the role of physiological features of rice cultivars on the quantity of methane emitted by them has been presented. All plant parts—leaves, culm/stem and roots—can play a part. The rate of photosynthesis occurring in the cultivar and the extent of biomass generated as a result, also exert significant influence on methane emission.

Acknowledgements SAA thanks the Council of Scientific and Industrial Research (CSIR), New Delhi, for the Emeritus Scientist grant (21(1034)/16/EMR-II).

References

Abbasi T, Abbasi SA (2010) Renewable energy sources: their impact on global warming and pollution. PHI Larning, New Delhi, pp x + 306

Abbasi T, Abbasi SA (2011) Decarbonization of fossil fuels as a strategy to control global warming. Renew Sustain Energy Rev 15(4):1828–1834

Abbasi T, Gajalakshmi S, Abbasi SA (2009) Towards modeling and design of vermicomposting systems: mechanisms of composting/vermicomposting and their implications. Indian J Biotech 8:177–182

Abbasi SA, Nayeem-Shah M, Abbasi T (2015) Vermicomposting of phytomass: limitations of the past approaches and the emerging directions. J Clea Prod 93:103–114

Abbasi SA, Dhanuja C, Abbasi T (2019) Emission of greenhouse gases from Indian wetlands: an overview. In: Proceedings of the National Academy of Sciences, India Section B. Under review (in press)

Aulakh MS, Bodenbender J, Wassmann R, Rennenberg H (2000) Methane transport capacity of rice plants. II. Variations among different rice cultivars and relationship with morphological characteristics. Nutr Cycl Agroecosys 58:367–377

Aulakh MS, Wassmann R, Bueno C, Kreuzwieser J, Rennenberg H (2001) Characterization of root exudates at different growth stages of ten rice (Oryza sativa L.) cultivars. Plant Biol 3:139–148

Aulakh MS, Wassman R, Rennenberg H (2002) Methane transport capacity of twenty-two rice cultivars. Agric Ecosyst Environ 91:5971

Baruah KK, Gogoi N, Gogoi B, Goswami S, Barman B, Gupta PK (2002) Plant and soil factors associated with methane emission from irrigated rice ecosystem of Assam. In: Van Ham J, Baede APM, Guicherit R, Williams-Jacobse JGFM (eds) Non-CO_2 greenhouse gases scientific understanding, control options and policy aspects. Millpress-Rotterdam, The Netherlands, pp 101–116

Baruah KK, Gogoi B, Gogoi P, Gupta PK (2010) N_2O emission in relation to plant and soil properties and yield of rice varieties

Bharali A, Baruah KK, Gogoi N (2017a) Methane emission from irrigated rice ecosystem: relationship with carbon fixation, partitioning and soil carbon storage. Paddy Water Environ 15(2)

Bharali A, Baruah KK, Gogoi N (2017b) Potential option for mitigating methane emission from tropical paddy rice through selection of suitable rice varieties. Crop Pasture Sci 68(5):421–433

Butterbach-Bahl K, Kahl M, Mykhayliv L, Werner C, Kiese R, Li C (2009) A European-wide inventory of soil NO emissions using the biogeochemical models DNDC/Forest-DNDC. Atmos Environ 43(7):1392–1402

Dhanuja C, Saxena DK, Abbasi T, Abbasi SA (2019) Effect of application of vermicompost on methane emission and grain yield of Chinna Ponni paddy crop. Paddy Water Environ 17(4):797–802

Dhanuja C, Abbasi T, Abbasi SA (2020) Fertilization of paddy cultivation with vermicompost: a critical mini review, Organic Agriculture (in press)

Das K, Baruah KK (2008a) Methane emission associated with anatomical and morphological characteristics of rice (Oryza sativa L.) plant. Physiol Plant 134:303–312

Das K, Baruah KK (2008b) A comparison of growth and photosynthetic characteristics of two improved rice cultivars on methane emission from rainfed agroecosystem of northeast India. Agr Ecosyst Environ 124(1–2):105–113

Denier van der Gon HAC, Kropff MJ, van Breemen N, Wassmann R, Lantin RS, Aduna E, Corton TM, van Laar HH (2002) Optimizing grain yields reduces CH_4 emission from rice paddy fields. Proc Natl Acad Sci USA 99:12021–12024

Ferch NJ, Romheld V (2001) Release of water-dissolved nitrous oxide by plants: does the transpiration water flow contribute to the emission of dissolved N2O by sunflower. In: Horst WJ et al (eds) Plant nutrition—food security and sustainability of agro-ecosystem. Kluwer Academic Publishers, Netherland, pp 228–229

FAOSTAT (2009) Agriculture organization of the United Nations. Statistical database

Ganesh PS, Ramasamy EV, Gajalakshmi S, Abbasi SA (2005) Extraction of volatile fatty acids (VFAs) from water hyacinth using inexpensive contraptions, and the use of the VFAs as feed supplement in conventional biogas digesters with concomitant final disposal of water hyacinth as vermicompost. Biochem Eng J 27(1):17–23

Gogoi N, Baruah KK, Gogoi B, Gupta PK (2003) Physiological parameters of rice associated with emission of methane from different agroecosystem. Ind J Plant Physiol (Special Issue) 597–601

Gogoi N, Baruah KK, Gupta PK (2008) Selection of rice genotypes for lower methane emission. Agron Sustain Dev 28(2):181–186

Jimenez J, Lal R (2006) Mechanism of sequestration in soils of Latin America. Crit Rev Plant Sci 25:337–365

Kerdchoechuen O (2005) Methane emission in four rice varieties as related to sugars and organic acids of roots and root exudates and biomass yield. Agr Ecosyst Environ 108(2):155–163

Kimura M, Murase J, Lu Y (2004) Carbon cycling in rice field ecosystems in the context of input, decomposition and translocation of organic materials and the fates of their end products (CO_2 and CH_4). Soil Biol Biochem 36(9):1399–1416

Leakey ADB, Ainsworth EA, Bernacchi CJ, Rogers A, Long SP, Ort DR (2009) Elevated CO_2 effects on plant carbon, nitrogen, and water relations: six important lessons from FACE. J Exp Bot 60:2859–2876

Luithui C, Dhanuja C, Abbasi T, Abbasi SA (2017) Greenhouse gas emissions and wetlands: a brief review of the state-of-the-art. Int J Environ Sci Technol 8(2):01–50

Nisbet RER, Fisher R, Nimmo RH, Bendall DS, Crill PM, Gallego-Sala AV, Hornibrook ERC, Lopez-Juez E, Lowry D, Nisbet PBR, Shuckburgh EF (2009) Emission of methane from plants. Proc R Soc Lond B Biol Sci 276(1660):1347–1354

Pangala SR, Gowing DJ, Hornibrook ER, Gauci V (2014) Controls on methane emissions from Alnus glutinosa saplings. New Phytol 201(3):887–896

Tabassum-Abbasi, Chirchom L, Abbasi T, Abbasi SA (2020a) Modelling methane and nitrous oxide emissions from rice paddy wetlands in India using artificial neural networks (ANNs), water, in press

Tabassum-Abbasi, Chirchom L, Abbasi T, Abbasi SA (2020b) A model to forecast methane emissions from topical and sub-tropical reservoirs on the basis of artificial neural networks, water (in press)

Tanaka Y, Sugano SS, Shimada T, Hara-Nishimura I (2013) Enhancement of leaf photosynthetic capacity through increased stomatal density in Arabidopsis. New Phytol 198:757–764

Tanaka Y, Kumagai E, Tazoe Y, Adachi S, Homma K (2014) Leaf photosynthesis and its genetic improvement from the perspective of energy flow and CO_2 diffusion. Plant Production Science 17:111–123

Climate Change Impact on Tourism-Based Livelihood and Related Youth Migration—A Case Study for Nainital, Uttarakhand, India

Shivakshi Jasrotia and Madhuben Sharma

Abstract The current study focuses on the sustainable development of tourism in Uttarakhand state of India by understanding climate change and its impact on the local environment with relevance to both livelihood and tourism. This research communication also brings forth the change in the current situation of livelihood and related youth migration observed in the Uttarakhand state of India due to sinking tourism. One of the fastest and biggest growing sectors today is tourism which has also been noted as the fastest growing industry with significant environmental, cultural, social, and economic effects. With changing trend of the environment due to climate change impact, added vulnerabilities, more human interferences, and time and again occurring natural hazards, there has been a sharp decline in tourism in a few regions of Uttarakhand state. This has largely affected the younger people who have relied on tourism related livelihood options like boating, water sports, paragliding, and other aquatic sports activities for their daily earnings. Another relevant change observed in the last decade as per the survey and interview done is the migration of youth population to nearby state, especially New Delhi for better growth opportunities. Sustainable tourism is one of the major concerns of achieving sustainable development in Uttarakhand state. This primary aim of tourism development requires an understanding of climate change and its impact on environment and has a properly focused strategy for remediation of land or other alternatives for nullifying the impact. One of the major objectives is to investigate the status of the current issue and suggest a necessary plan for sustainable tourism development. It also focuses on the formation of active social groups for knowledge dissemination in masses about climate change trend. This study was done in the prime rural base of Uttarakhand state with tourism-prone location where previously rural residents and youth were mostly depending on tourism options for their livelihood. As a methodology, a

S. Jasrotia (✉)
Delhi Institute of Pharmaceutical Sciences and Research (DIPSAR), Pushp Vihar, New Delhi, India
e-mail: shivakshi_jasrotia@yahoo.com

M. Sharma
University of Petroleum and Energy Studies, Dehradun, Uttarakhand, India
e-mail: madhubensharma06@gmail.com

N. A. Siddiqui et al. (eds.), *Advances in Water Pollution Monitoring and Control*,
Springer Transactions in Civil and Environmental Engineering,
https://doi.org/10.1007/978-981-32-9956-6_8

field-based systematic survey was carried out based on the general questionnaire in the selected tourism destinations to find out the ongoing problems and issues. This research focuses on integrated approach and establishment/suggestions of new plans for coping with the changing climate impact for sustainable development of tourism in Uttarakhand. This would be of help to related government agencies and the statistics department of population control for their policy-making to cater to the shortfall of the lower number of youth availability in the village. The study concludes to synthesize all the findings and further gives feasible suggestions and recommendations for the development of tourism in this region.

Keywords Sustainable development · Youth migration · Tourism · Cultural · Environmental · Integrated approach

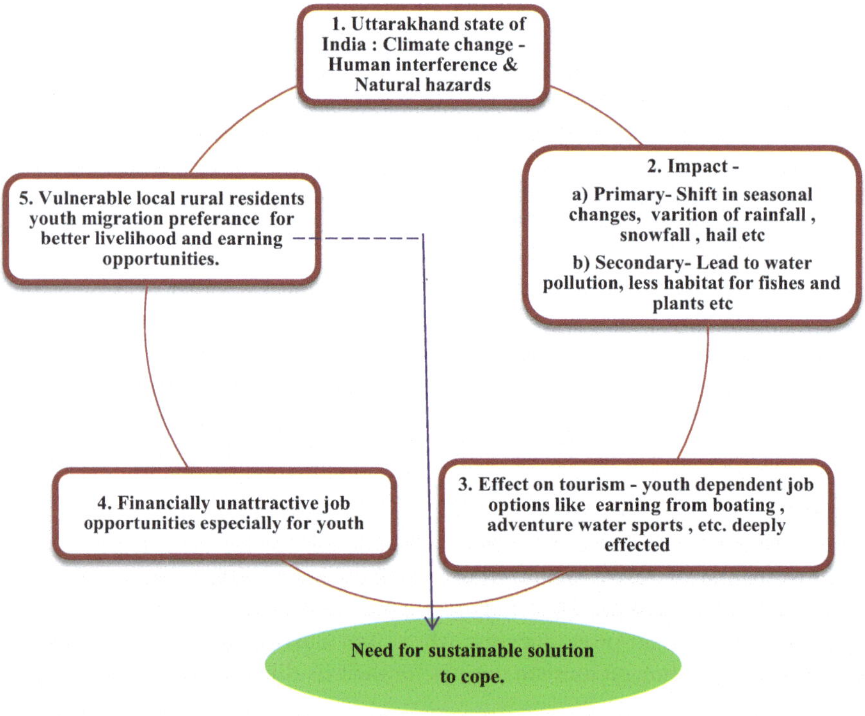

Schematic of research for youth migration to developing states for better job opportunities then coping with sinking Schematic tourism—An impact of climate change on tourism needing sustainable solutions

1 Introduction

Nainital is a beautiful city in Uttarakhand state of India and one of the main tourist attractions of India. It is nestled in Kumauni region of Uttarakhand, as one of the four beautiful Kumauni lakes, of India at an elevation of 2000 m. The specialty and focal point of the beautiful city lie with the naturally present lakes and scenic beauty (Fig. 1a, b). Reported since British time, this hill station is considered as a tourist paradise as it is set around Nainital lake with religious Hindu temple of Nanda Devi toward its north shore.

What further makes this lake one of the most pristine tourist locations especially during hot summers is its ready connectivity with India's capital, New Delhi. The total distance is nearly 300 km easily connected by railways and road. What this lake offers is options for earning income by alternatives like boating and parallel cable car Snow View Points (SVP's) at an observation point of 2,270 mt approximately for Nanda Devi, Uttarakhand's highest peak. The place not only has beautiful scenic beauty but it also has a wide variety of naturally growing aquatic plants, blooms which have been home to special variety of foreign fish like mirror carp (mahseer) and trout (Fig. 2). Species like *Schizothorax sinuatus*, *Schizothorax richardsoni,* and *Schizothorax plagiostornus* are commonly found in this lake all year around.

What was observed a decade ago and what is present today is a very different entity for this beautiful lake that is degrading with time. A look at the lake's hydrography reveals that on its northwest side is surrounded by Naina peak and has Tiffin Top on the southwest side. Furthermore, the snow peaks are visible on the front north side of this lake. Coniferous forest trees cover these hill ranges, with tropical monsoon climates having a maximum temperature ranging to only 24.6 °C in summers.

Fig. 1 **a** Nainital lake—famous tourism spot. **b** Scenic view of the lake (Picture taken on 20th October 2016)

Fig. 2 Fish species—Lake Nainital (Picture taken on 20th October 2016)

Maximum temperature of 25 °C and a minimum of 0.5 °C have been recorded with tropical monsoon climatic conditions. The range of Ph has been found to be between 8.4 and 9.3; this reportedly is to be of alkaline nature. A series of climate change effects and detracting flora and fauna, especially affecting the rich carp fish, have been witnessed by regular tourist visitors themselves.

With time, this lake and its natural biodiversity have been found to be in danger. Local resident and my bus driver Mr. Chandan have witnessed that there is less and at times "No" snowfall in winters as it used to be a decade ago. Also, the rainfall has gone scanty in the much needed winters for the balance of the environment. There has been a sharp decline in the water level with deteriorating water quality (Sharma 2014) affecting both flora and fauna and residents depending on the lake water. This is not just affecting the environment and lake biodiversity but also raising an alarming situation for the reduction in tourists, which is further kicking the local economy support of depending boatman and their families. Forced youth migration is the current scenario and solution opted where youngsters are avoiding to work as boating man and guide and prefer to fetch job opportunities in the nearest metro city, New Delhi. With time, the lake has come under the threat of climate change and lack of government-supported measures. The impact is straight on local fishermen, shopkeepers, and boatman depending family for their daily livelihood.

2 Study Area and Methodology

Figure 3 depicts the location of lake Nainital which geographically is located in the densely populated valley of the Kumaun Himalaya (Nainital district) of Uttarakhand. The lake is 1.5 km in length and is around 0.36 km in width appearing to be crescent in shape. Geographic location has its marked boundaries between 29220 N and 29240 N

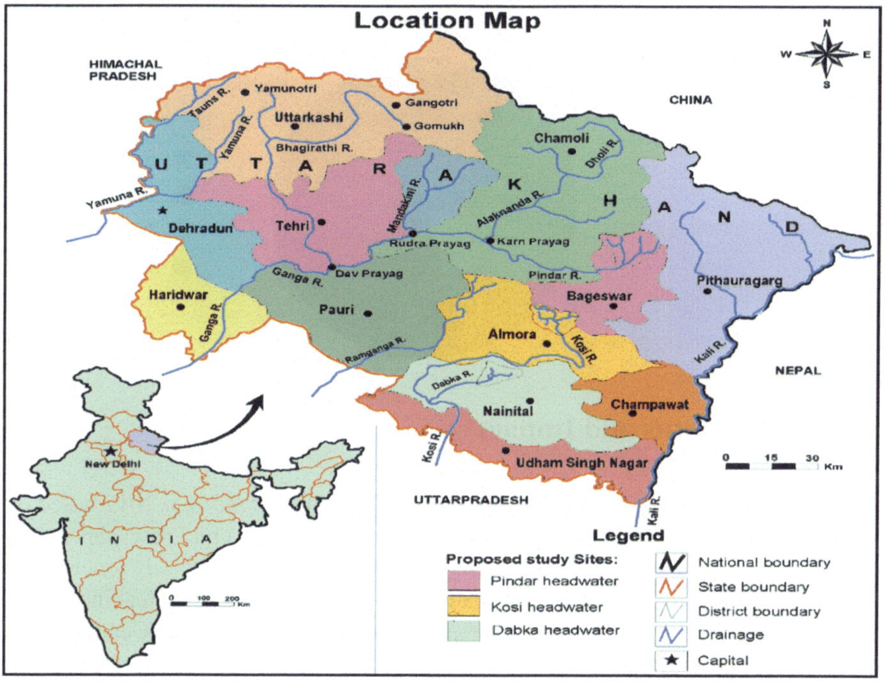

Fig. 3 Location of lake Nainital, Uttarakhand, India (*Source* Online Map Of India)

(latitude) and longitudes of 79260 1300 E and 79280 500 E (Choudhary et al. 2009). It covers a surface area of 48 ha. It shares proximity to the Main Boundary Thrust which also happens to be the partition separating the Siwaliks from Lesser Himalaya lake. Further, with very complex geology, the Nainital lake is bounded by the Sher-ka-Danda hill in the east, the landslide flat deposit in the north, the Ayarpatta hill in the west, and Balia ravine in the south (Singh et al. 2016).

A survey was done with the help of the local people with group discussions during my research study in the month of winters, October 2016 (chosen due to the peak season of tourist visit). Nearly 50 household people and shopkeepers were asked about their observance in terms of the lake and related climate change impact. A decadal comparison was made with relevant questions asked and data was recorded. The survey included a questionnaire with questions related to the affecting climate change factors with reference to lake water quality. The results have further been categorized in the form of impacts as per received data for questionnaire with the response to factors responsible for climate change (Fig. 4).

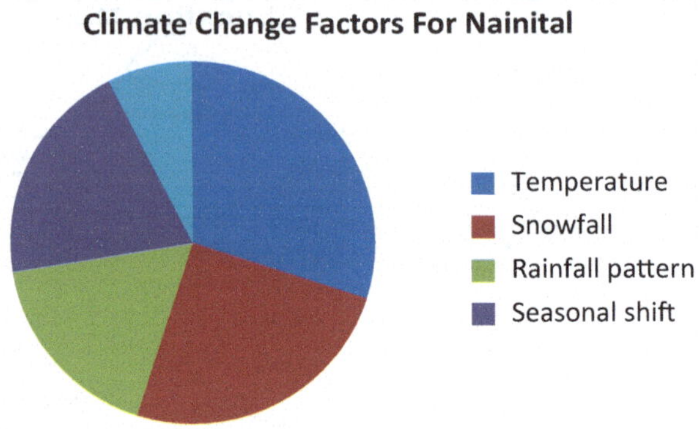

Fig. 4 Climate change drivers for lake Nainital study

3 Current Status and Impact

Table 1 gives an overview of the current status of the lake for the year 2017 based on the survey and recent literature reports. It has been reported by chief newspapers that both the huge lakes of Nainital will go dry (Fig. 5). The lake has already touched the abysmal level of 14 feet below normal water level which marks a very clear indication of an upcoming and impending ecological disaster. According to the Centre's draft State of the Environment report, 2015, the loss to the Indian economy from water resource depletion is pegged at 6% by 2050 if the country continues to manage its water the same way (Hindustan time's e-paper, 2017). Also, active research has already shown the fate of these lakes with miserable water quality both for flora and fauna (Sharma 2014).

Figure 6 depicts the linkage between climate change and vulnerability to agriculture and tourist attraction with related youth migration. Related discussion is present below.

3.1 **Water quality and water shrink**: Quality of water is an issue but when the quality of available water is affected, it has a major impact on the health of rural residents, flora, and fauna. Research study shows that the water quality in terms of salinity and hardness has increased with the lake shrinking and drying up. This has shown a sharp fall in the number of carps and trout's breeding period. Also, aquatic plant species have now stopped blooming in their regular time frame due to change in cycles of monsoon and seasonal variation.

3.2 **Agriculture Impact**: This change in climate (monsoon shower cycles) has also brought forth the issue of crops of winters being affected. Barley and wheat growth with the disturbance in their cycle of germination have been adversely affected (Singh 2016). This is another reason for the reduction in the livelihood earning which was directly dependent on agriculture practices for long, carried forward from ancestral land. Depending on agriculture and horticulture (honey making and jam making) have been an important way of Uttarakhand people's earning for the family. In reference to the impact and outcomes, apple production has declined in its good harvest which has been a setback to state's apple business

Table 1 Shows the nature and kind of impact of climate stressors on Nainital Lake and nearby surroundings

Climate stressors	Frequency and variation level	Interference/sensitivity	Nature of impact	Need of sustainable solution
Increasing temperature	High	Decreased the amount of lake water in basin area catchment. Water quality more silted Forest fires	1. Yes, boating cycle is affected 2. Tourist prefers less mobility with increased temperatures[a] 3. Youth migration and more labor demand	Yes, ecotourism is needed Alternatives to job security and engagement of youth are to be kept at priority. Effective government policies need to be formulated
Decreasing snow	High	Erratic and less	1. Sports acuities like paragliding, mountain climbing affected 2. Less tourist attraction 3. Youth migration[b]	Yes, ecotourism is needed Alternative to job security and engagement of youth
Rainfall patterns	Moderate to high	Level of water decreases, mudslides are very common	1. Disturbance in scenic beauty and boating cycle 2. Mudslides block tourist	Yes, ecotourism is needed as destination becomes less attractive

(continued)

Table 1 (continued)

Climate stressors	Frequency and variation level	Interference/sensitivity	Nature of impact	Need of sustainable solution
Seasonal shift	Very high	Disturbed the natural flora, fauna, and fish species breeding cycle. Impact on agriculture	1. Breeding cycle of carps and other trout's has been disturbed 2. Depleting lake and vegetation[d] 3. Direct effect on fish catch (restricted to limited area) 4. Germination rates have been deeply affected for important crops like rice, paddy, etc.	Yes, ecotourism is needed Lake restoration initiatives to be increased
Cloud burst and forest fire	Rare, but strong	Blocks road and transport	1. Sharp dip in the number of tourist 2. Disturbance to ecological environment and forest 3. Economy loss to state due to less tourist visit and decrease in hiking and camping[c]	Yes, ecotourism is needed

[a]Bhakuni (2017)
[b]Singh et al. (2016)
[c]Mamgin and Reddy (2017)
[d]Sharma (2014)

Fig. 5 Deteriorating lake (*Source* Hindustan times News report, May 2017)

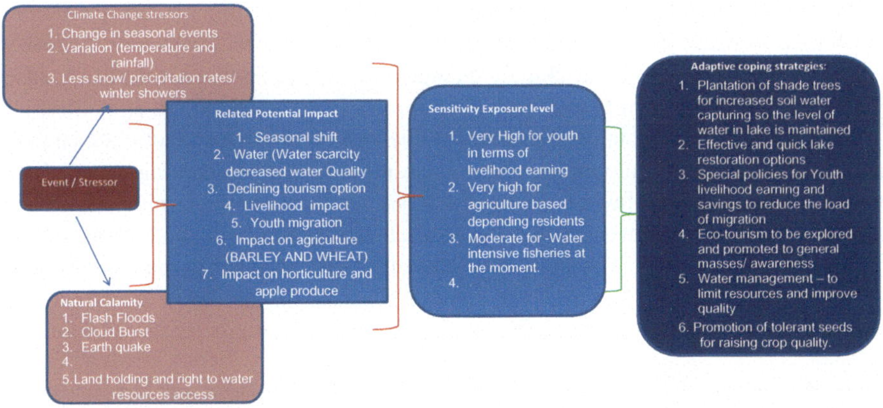

Fig. 6 Linkage between climate change—vulnerability to agriculture and tourist attraction with related youth migration

and economy. Tourists with time had aided in the state's apple purchase and related business but, the numbers have depreciated quickly.

3.3 **Environmental impact on tourism/youth migration**: Massive effect has been witnessed for the sharp decline in the number of tourists approaching Nainital lake for summer vacations. This has disturbed the natural and the inherited ways of earning from natural sources like lake water boating, paragliding, mountain trekking, etc. The most critical impact has been the migration of the youth with respect to not trying to get an economy hike and earning options. This is due to the drying lakes and disturbed seasons largely affecting tourism. A majority of the youngsters today is found less interested in inherited practices of earning from natural sources like water sports activities, boating, fishing, and paragliding. It has been witnessed that over time in Nainital, there is lack of opportunities, lack of ease of using these opportunities to get this financial security and earning options, less revenue from agriculture, and decreasing tourist masses and numbers. In response to the climate generated stresses and, unpredictable weather changes, the population on its own, especially the youth, move to other areas where opportunities of making a livelihood are easily available (Mamgain and Reddy 2017).

4 Conclusion and Discussion

Migration is a self-explanatory word which denotes there is a change involved and a response has been adapted. In this small case study, a vision for the current status of natural wetlands and lakes with withering surroundings has been highlighted. India has thousands of traditional wetlands in the form of ponds, lakes, vayalgam, ahars, bawdis, and talab. But with growing climate change, its impacts, and economic growth, these water bodies are under stress. In the recent past, lake water is greatly deteriorated due to the global climate change and anthropogenic activities because they affect both the quantity and quality of the available water (Singh et al. 2016).

The lake is definitely under some threat due to a combination of climate change and poorly handled development. There is an urgent need of adapting to solutions which are sustained and favor tourism as that is a needed revenue-earning alternative in this region. Ecotourism could serve as an alternative for favoring both earnings and saving of natural beings. The study thus puts efforts to sustainably manage Nainital and its surroundings so that even the coming generations can enjoy the essence of nature while the region can serve the tourists and guests with its preserved beauty.

Acknowledgements The authors are grateful to the local village residents and their families for their extended cooperation.

References

Bhakuni K (2017) Issues and challenges of sustainable tourism in Nainital, Uttarakhand. In: Sharma P, Rajput S (eds) Sustainable smart cities in India. The urban book series. Springer

Choudhary P, Routh J, Chakrapani G, Kumar B (2009) Biogeochemical records of paleoenvironmental changes in Nainital Lake, Kumaun Himalayas, India. J Paleolimnol 42:571–586

Mamgain RP, Reddy DN (2017) Out-migration from the hill region of Uttarakhand: magnitude, challenges, and policy options. In: Reddy D, Sarap K (eds) Rural labour mobility in times of structural transformation. Palgrave Macmillan, Singapore

Sharma M (2014) Water quality assessment of the central Himalayan Lake, Nainital. Adv Environ Chem

Singh N (2016) Climate change and environment issues. The Energy and Resources Institute (TERI)
Singh D, Rai SP, Kumar B et al (2016) Environ Earth Sci 75:1380. https://doi.org/10.1007/s12665-016-6177-1
http://www.hindustantimes.com/opinion/chandigarh-s-sukhna-and-nainital-s-naini-lakes-are-going-dry-here-s-why/story-qaIYNhFo298jLlC6Hx5bRK.html. Accessed on 15/january/2018

Study of Synthesis and Biological Importance of 1,4-Dihydropyridines Derivatives

Neetu Tripathi

Abstract 1,4-Dihydropyridines (1,4-DHPs) are very important heterocyclic structures from which a large number of compounds are known in the field of drugs and pharmaceuticals. Interest in 1,4-dihydropyridines is mainly due to nicotinamide adenine dinucleotide (NADH), a coenzyme, with unique ability to reduce various functional groups in biological systems. They are also known for resistance reversal properties and antitubercular activities. The present study involves the reaction of three components like an aldehyde, enamine, and di keto compounds to synthesize 1,4-dihydropyridines in good yields by using green chemistry idea that is the use of an eco-friendly solvent and a catalyst TBAHS.

Keywords 1,4-Dihydropyridine · Catalyst · Aldehydes · Enamines · β-keto compound

1 Introduction

Dihydropyridines (DHPs) are the heterocyclic compounds which contain nitrogen as a hetero atom. 1,4-dihydropyridines are very useful compounds with a feasible heterocyclic ring with different substitutions at several positions. They are also known as cognition enhancers (Cooper et al. 1992). Some 1,4-DHP compounds are reported as antioxidants and drug resistance modifiers (Shridhar and Perumal 2005). These molecules are known for various biological activities like calcium channel blockers (Triggle 2003). They are also known as antihypertensive (Bossert et al. 1981; Nakayama and Kasoka 1996), antitumor (Saushins and Duburs 1988; Mager et al. 1992), stress protective (Tarasenko et al. 2002), anticonvulsant (Tusell et al. 1993), analgesic, anti-inflammatory (Sushilkumar and Devanand 2002), and antidiabetic (Saushins and Duburs 1988; Manhold et al. 1992; Klusa 1995; Bretzel et al. 1993; Boer and Gekker 1995; Bretzel et al. 1992) agents. It was reported that 1,4-DHPs are also known as antitubercular agents (Kharkar et al. 2002). The presence of lipophilic

N. Tripathi (✉)
Department of Chemistry, MKP P.G. College, Dehradun, India
e-mail: neetu_tiwari78@yahoo.com

N. A. Siddiqui et al. (eds.), *Advances in Water Pollution Monitoring and Control*, Springer Transactions in Civil and Environmental Engineering,
https://doi.org/10.1007/978-981-32-9956-6_9

groups in substituted 1,4-dihydropyridines have been reported to have a very good anti-TB activity which makes them an emerging class of antitubercular agents in the area of drug discovery (Trividi et al. 2011). Their role as drugs for the treatment of various diseases makes them act as multifunctional lead molecules in drug discovery. They are well-known as neuroprotectants, antiplatelet aggregators and also are important in Alzheimer's disease as anti-ischemic agents (Klusa 1995; Bretzel et al. 1993). The resemblance of 1,4-dihydropyridines with nicotinamide adenine dinucleotide (NADH), a coenzyme involved in hydrogen transfer reactions in biological processes to reduce many functional groups in biological systems attracts the attention of medicinal chemists working in the area of drug discovery.

Recently, substituted 1,4-dihydropyridines have been reported as potent antitubercular agents (Srinivasa Rao and Laxmi 2013; Olejnikova et al. 2014) against *Mycobacterium tuberculosis* as well as *Mycobacterium smegmatis* and also possess antimicrobial activity against *Bacillus subtilis*, *E. coli*, and *Protius vulgaris* bacteria. Recently, a series of symmetrical, asymmetrical, and unsubstituted 1,4-dihydropyridines have been studied for antitubercular activity and one compound has been reported with antitubercular activity almost equal to that of pyrazinamide; familiar TB drug having pyridine nucleus (Swarnalatha et al. 2011). Tuberculosis is an infectious disease which is the most important cause of death in India (Yajnik et al. 2002). TB is caused by slow-growing bacterium *Mycobacterium tuberculosis*. It is a respiratory transmitted disease and it affects nearly 32% of the world's population and in India, nearly one-third population has been affected. About 1.8 million persons develop the disease in India and 8,00,000 are having latent tuberculosis which is non-symptomatic. Further emergence of MDR-TB (Multi-drug-resistant TB) worldwide worsened the situation. It is revealed almost all over the world for tuberculosis that the current therapeutic measures are not sufficient; therefore, many pharma industries and research institutes are engaged in the research of finding new antitubercular regimens by designing and synthesizing various compounds. There are many of the existing drugs or compounds like Amlodipine, Nifedipine, Nitrendipine, Nimodipine, Lacidipine, and Isradipine, etc., in clinical trials that possess this dihydropyridine skeleton as shown in Fig. 1.

Therefore, the development of new drugs for the treatment of multi-drug-resistant tuberculosis (MDR-TB) is the top priority area worldwide (Rozwarski et al. 1998). In India, TB kills two persons in every 3 min and 1000 in a day. Approximately 90% of patients having HIV and TB, both, did not survive for a longer time after clinical diagnosis as per WHO reports. 1,4-dihydropyridines are known for their resistance reversal properties (Mdluli et al. 1998) and antitubercular activities (Manvar et al. 2010) Based on our earlier research, since sugar molecules are capable of showing a lot of structural variation, some sugar-based substituted dihydropyridines (Tewari et al. 2004; Tripathi et al. 2002) are thought to be synthesized and their bioevaluation will be carried out for antimicrobial activities particularly antitubercular activity against *M. tuberculosis* and antifungal activity. Since the sugar derivatives (Chennat and Eisner 1975) are known in molecular recognition properties for biological activities, the bacterial cell wall is taken as a selective target for drug discovery.

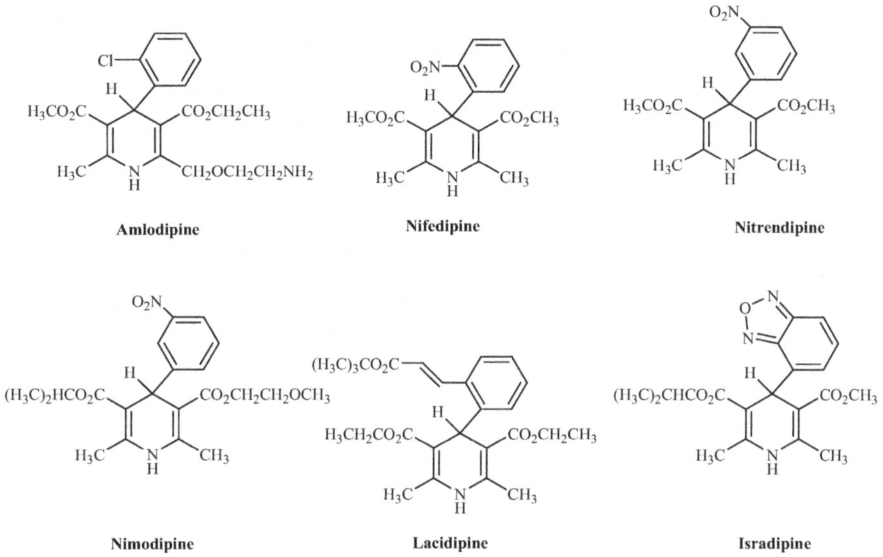

Fig. 1 Dihydropyridines based drugs or candidate drugs

2 Rationale for Synthesis and Methodology

The 1,4-DHP ring has been synthesized by a number of different methods like condensation of β-keto esters (van Bergen and Kellogg 1976) and using crown ether integrated with dihydropyridines as natural coenzyme NADH mimics (Kellogg et al. 1980; Ganesh Kumar et al. 2013). To generate green chemistry idea in the synthesis of substituted 1,4-dihydropyridines, we have developed a method in which eco-friendly solvent and catalyst conditions have been optimized under microwave radiation. Our work involves synthesis of 2,3,5,6 tetrasubstituted-4-glycosyl-1,4-dihydropyridines using aldehydes, enamines, and β-keto compound in the presence of a suitable catalyst by optimizing solvent conditions. My objective is to design sugar-substituted dihydropyridines because the carbohydrates are well-known for their pharmacokinetic and pharmacodynamic properties in the biological system due to good solubility. We have taken carbohydrate as a starting material for the preparation of one of the synthones, i.e., glycosyl aldehydes and nontoxic nature of sugars make it vulnerable for the synthesis of 2,3,5,6 tetrasubstituted-4-glycosyl-1,4-dihydropyridines by optimizing Hantzch synthesis using a suitable catalyst and eco-friendly conditions. Series of compounds are planned to be synthesized to see the effect of various substituents in biological activities. Once the 2,3,5,6 tetrasubstituted-4-glycosyl-1,4-dihydropyridines are synthesized, then it could be explored further because 1,4-dihydropyridine skeleton is also a good precursor of the corresponding substituted pyridine derivative and also acts as a good source of hydride for reduction propose.

Since carbohydrate are nontoxic and well-known for their structural variation properties and they are extremely cheap, abundant, and act as chiral auxiliaries, synthesis of 2,3,5,6 tetrasubstituted-4-glycosyl-1,4-dihydropyridines has been carried out by optimizing Hantzch synthesis using a suitable catalyst and eco-friendly conditions and similarly some other compounds with aromatic substituent at 4-position would also be synthesized. First, glass apparatus were dried and solvents were evaporated under reduced pressure. Thin-layer chromatography has been performed in order to monitor the progress of the reaction using silica gel 60 F_{254} plates. For the detection of spots to check the reaction with time, iodine vapors were used and 5% sulphuric acid in ethanol was used as a spraying agent followed by heating at 100 °C, or by spraying with Dragendorf reagent. Silica gel (60–120 mesh) was used for column chromatography. UV spectrophotometer is also thought to be utilized for monitoring the reactions. Both pyranoses and furanoses forms of sugars like glucose and galactose would be converted to respective aldehydes (**1**) by the known strategy of protection, selective deprotection, and oxidation of hydroxy compounds of sugars. Then as an alternate methodology using one equivalent of aldehydes (**1**), one equivalent enamines (**2**) of the β-keto compound which were thought to be involved in 1,4-dihydropyridine synthesis and one equivalent of the β-keto ester or diketone (**3**) in the presence of a suitable catalyst in eco-friendly solvent would lead to the formation of the desired 2,3,5,6, tetrasubstituted-4-glycosyl-1,4-dihydropyridine compounds in better yield as shown in Fig. 2.

Based on our earlier observations, the phase transfer catalysts like Tetra Butyl Ammonium Hydrogen Sulfate (TBAHS) work well in such cyclization reactions (Tewari et al. 2003, 2004; Tripathi et al. 2002). Our method involves the use of less expensive and easily available phase transfer catalyst TBAHS and diethylene glycol as a green solvent to synthesize various kind of 4-substituted dihydropyridines having similar or different substituent. The reaction is also good with aromatic and heteroaromatic aldehydes. We have also optimized the reaction with catalyst solvent-free conditions under microwave radiation to give the compound in optimum yield.

As sugars are known to induce asymmetry during organic synthesis, it was of considerable importance to see their effect here too. The purpose could be achieved by taking enamine of a β-keto compound different from the β-keto compound used as one of the substrates in the above reaction. Thus, condensation of an enamine which was prepared from methyl acetoacetate and ammonium acetate as usual, with

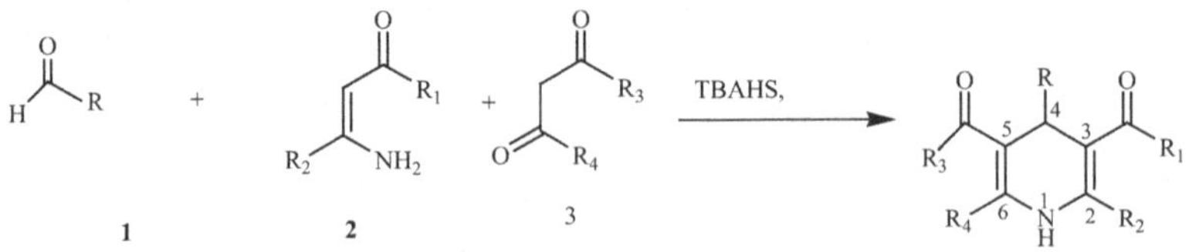

R= Sugar/aryl R_1/R_2=CH_3 or OCH_3 R_3/R_4=CH_3 or OCH_3

Fig. 2 Three components reaction-in the presence of catalyst TBAHS and ethylene glycol

glycosyl aldehydes with 3-*O* benzyl and acetyl acetone separately led to the formation of respective glycosyl dihydropyridines with 3-*O* benzyl substituent in good yields.

Similarly, the reactions of aromatic aldehydes (3-nitro benzaldehyde and 3-pyridyl aldehyde) were reacted with the above enamines separately with acetyl acetone and ethyl acetoacetate leading to the formation of respective 4-aryl-dihydropyridines in quantitative yields. The structure of all these compounds has been determined on the basis of their spectroscopic data and analysis. All the above compounds showed $[M + H]^+$ peak in their MS FAB spectra. In IR spectra of these compounds, the characteristic stretching absorption bands have been observed at around 3432, 1628–1692 cm^{-1} corresponding to NH and $C = O$ of ketone or ester functionality. The data of the structural analysis was obtained with the help of CDRI, Lucknow. The bioevaluation of the compounds for antimicrobial activity would be done by screening them against the bacterial and fungal strains. Bacterial cell wall polymers like mycolic acid and peptidoglycans have been chosen as a selective target to develop new antimicrobial agents.

3 Conclusions

This study is important not only from the basic research point of view but also for medicinal significance. The green synthetic process is pollution-free and eco-friendly as it reduces toxic chemicals. The prevalence of infectious diseases and the emergence of drug resistance (Ganesh Kumar et al. 2013) address that the development of new antimicrobial agents is still our need. 1,4-dihydropyridine derivatives have pharmacological importance; therefore, their synthesis is a matter of research and these molecules also have many applications in building highly efficient antibiotics, anti-inflammatory, antidiabetic, antitumor, and antitubercular medicines which are still a great challenge in the field of drug discovery.

References

Boer R, Gekker V (1995) Drugs Fut 20:499
Bossert F, Meyer H, Wehinger E (1981) Angew Chem Int Ed Engl 20:762
Bretzel RG, Bollen CC, Maeser Maeser E, Fedrlin KF (1992) Am J Kidney Dis 17:465
Bretzel RG, Bollen CC, Maeser E, Fedrlin KF (1993a) Am J Kidney Dis 21:53
Bretzel RG, Bollen CC, Maeser E, Fedrlin KF (1993b) Am J Kidney Dis 21:53
Chennat T, Eisner V (1975) J Chem Soc Perkin Trans 1:926
Cooper K, Fray MJ, Parry MJ, Richardson KK, Steele J (1992) J Med Chem 35:3115
Ganesh Kumar S, Adithan C, Harish BN, Sujatha S, Roy G, Malini A (2013) J Nat Sci Biol Med 4(2):286–291
Kellogg RM, van Bergen TJ, van Doren H, Hedstrand D, Kool J, Kruizinga W, Troostwijk CB (1980) J Org Chem 45:2854
Kharkar PS, Desai B, Gaveria H, Varu B, Loria R, Maliapara Y, Shah A, Kulkarni VM (2002) J Med Chem 45:4858

Klusa V (1995a) Drugs Fut 20:135
Klusa V (1995b) Drugs Fut 20:135
Mager PP, Coburn RA, Solo AJ, Triggle DJ, Rothe H (1992) Drug Des Discov 8:273
Manhold R, Jablonka B, Voigdt W, Schoenfinger KF, Schravan K (1992) Eur J Med Chem 27:229
Manvar AT, Raghuvir RS, Pissurlenkar (2010) Mol Divers 14:285–305
Mdluli K, Slayben RA, Zhu YQ, Ramaswamy S, Pan X, Mead D, Crane DD, Musser JM, Barry CE (1998) Science 280:1607
Nakayama H, Kasoka Y (1996) Heterocycles 42:901
Olejnikova P, Svorc L, Olsovska D, Panakova A, Vihonska Z, Kovaryova K, Marchalin S (2014) Sci Pharm 82:221–232
Rozwarski DA, Gregori AG, Barton DHR Jacobs, Jr WR, Sacchettini JC (1998) Science 279:98
Saushins A, Duburs G (1988a) Heterocycles 27:269
Saushins A, Duburs G (1988b) Heterocycles 27:269
Shridhar R, Perumal P (2005) Tetrahedron 61:2465
Srinivasa Rao N, Laxmi KL (2013) Sci Rev Chem Comm 3:141–149
Sushilkumar B, Devanand S (2002) Acta Pharmaceutica (Zagreb) A 54(4):281–287
Swarnalatha G, Prasanthi G, Sirisha N, Madhusudhana Chetty C (2011) Int J Chem Tech Res 3(1):75–89
Tarasenko LM, Neporada KS, Klusha V (2002) Bull Exp Biol Med 133(4):426–428
Tewari N, Katiyar D, Tiwari VK, Tripathi RP (2003) Tetrahedron Lett 44:6639–6642
Tewari N, Dwivedi N, Tripathi RP (2004) Tetrahedron Lett 45:9011–9014
Triggle DJ (2003) 1,4-Dihydropyridines as calcium channel ligands and privileged structures. Cell Mol Neurobiol 23(3):293–303
Tripathi RP, Tripathi R, Tiwari VK, Bala M, Sinha S, Srivastava A, Srivastava R, Srivastava BS(2002) Eur J Med Chem 37:773–781
Trividi AR, Dodia DK, Dholaria BH, Kataria VB, Bhuva VR, Shah VH (2011) Biorgan Med Chem Lett 21:5181–5183
Tusell JM, Barron S, Seratosa J (1993) Brain Res 622:99–104
van Bergen TJ, Kellogg RMJ (1976) Chem Soc Chem Commun 1976:964
Yajnik K, Chakraborty A, Jochem K (2002) New Delhi, The World Bank

Assessment of the Ability of a Novel Machine to Separate Different Types of Substrate–Earthworm–Vermicompost Mixtures

Tasneem Abbasi, S. M. Tauseef, T. Aravindan and S. A. Abbasi

Abstract A novel substrate–earthworm–vermiocompost separation (SEVS) machine is described which enables simultaneous, effective, clean, and rapid separation of finished vermicompost (vermicast), uncomposted substrate, and earthworms. The ability of the system was extensively tested with vermicomposted batches of paper waste, cow dung, and weeds—salvinia (*Salvinia molesta*), ipomoea (*Ipomoea carnea*), and lantana (*Lantana camara*). The effect of different batch sizes and also moisture content of the vermicast on the separation efficiency, with or without secondary separation, were tested. It is seen that SEVS machine achieves clean separation at all times and its precision is enhanced with the extent of pre-drying of the vermicast in the batch.

Keywords Vermicompost · High-rate vermireactors · SEVS machine · Separation of vermicast

1 Introduction

As has been brought out in Chap. 3 of this volume, billions of tons of waste biomass—termed "phytowaste" by us—of which no environmentally sound means of utilization has existed so far (Abbasi and Abbasi 2010a; 2010b) is likely to be utilized in the future. This is set to be made possible by the paradigm of "high-rate vermicomposting" and associated technology developed by Abbasi and co-workers (Abbasi et al. 2009, 2011, 2014, 2015; Tuaseef et al. 2013, 2014) which has made it possible to vermicompost phytowaste without the need of any manure supplementation or preprocessing. Even more significantly, the vermicomposting is achievable at a rate several

T. Abbasi (✉) · T. Aravindan · S. A. Abbasi
Centre for Pollution Control & Environmental Engineering, Pondicherry University, Chinnakalapet, Puducherry 605014, India
e-mail: Tasneem.abbasi@gmail.com

S. M. Tauseef
Department of Health, Safety and Environmental Engineering, University of Petroleum and Energy Studies, Dehradun 248007, India

N. A. Siddiqui et al. (eds.), *Advances in Water Pollution Monitoring and Control*, Springer Transactions in Civil and Environmental Engineering,
https://doi.org/10.1007/978-981-32-9956-6_10

times faster than that attained by the preexisting vermicomposting systems (which principally handle animal manure and are unable to process phytowaste). Backward integration of this step with extracting energy precursors (Ganesh et al. 2005; Abbasi and Abbasi 2011), generating nanoparticles (Ganaie et al. 2014; Anuradha et al. 2015; Mavukkandy et al. 2016), etc, and forward integration with agriculture (Hussain et al. 2015) are expected to further boost the applicability of this technology. It would also act as a bulwark against the risk posed by the highly colonizing weeds like salvinia (Abbasi and Nipany 1986; Khan and Abbasi 1998, 1999).

With vermicomposting likely to be done more commonly and at a much larger scale, it is necessary to pay attention to the harvesting of vermicast from the vermireactors along with the nondestructive separation of earthworms. This is particularly relevant to high-rate vermireactors because they need harvesting of vermicast and introduction of fresh feed much more frequently than is needed in the slower conventional systems.

Separation of vermicast from the uncomposted substrate and controlled removal of earthworms constitutes a critical component of the vermicomposting operation because it involves clean removal of the finished vermicompost from the vermireactor, together with the separation of the earthworms in such a way that they are not harmed. The system should also be able to quantitatively retrieve the undigested substrate so that it can be recycled. Moreover, as stated above, in order to obtain maximum benefit from the vermicomposting systems the vermicast is to be harvested at periodic intervals so that fresh feed can continue to be introduced. All this calls for an efficient vermicast–earthworm–substrate separation system.

As reviewed by Banupriya (2013) and Aravindan (2014), several mechanized harvesting equipment have been used but they can separate only either vermicast and substrate or substrate and earthworm. The systems are also energy-intensive in their operation. In order to facilitate the separation of vermicast, earthworm, and undigested substrate simultaneously, we have developed a machine called substrate–earthworm–vermicast separator (SEVS). This SEVS system is the first of its kind which enables the separation of earthworm, vermicast, and substrate simultaneously (Tauseef et al. 2014). We present here a refined version of SEVS and the results of extensive testing of its performance with several different substrate–earthworm–vermicompost batches.

2 The Vermicast–Earthworm–Substrate Separator (SEVS) Machine

The key elements of the SEVS machine are a light source, a shaker, and a set of sieves used to separate undigested substrate, earthworms, and vermicast from their mix in the vermireactor. Two of the embodiments of the SEVS system are shown in Figs. 1 and 2. The system consists of three modules arranged vertically, a supporting frame, a cam, a pulley, and a motor. A mild steel plate is fixed in the backside of the first and the second modules which is attached to a connecting rod. The spindle is

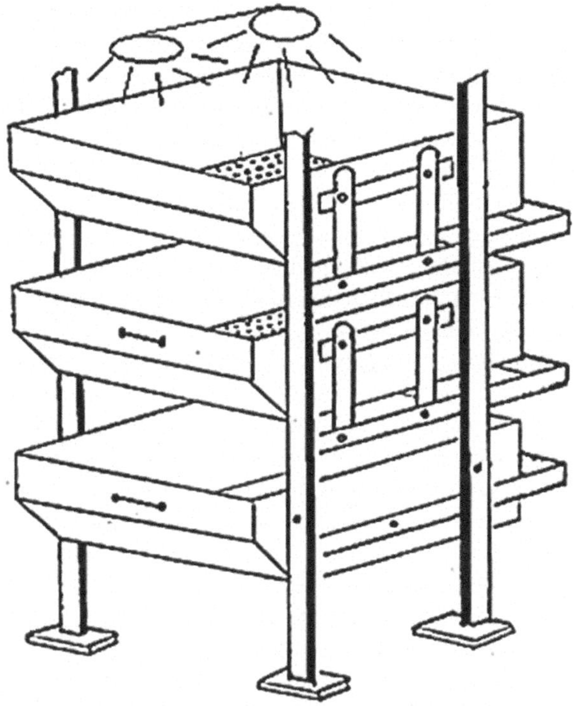

Fig. 1 An embodiment of a VESS system

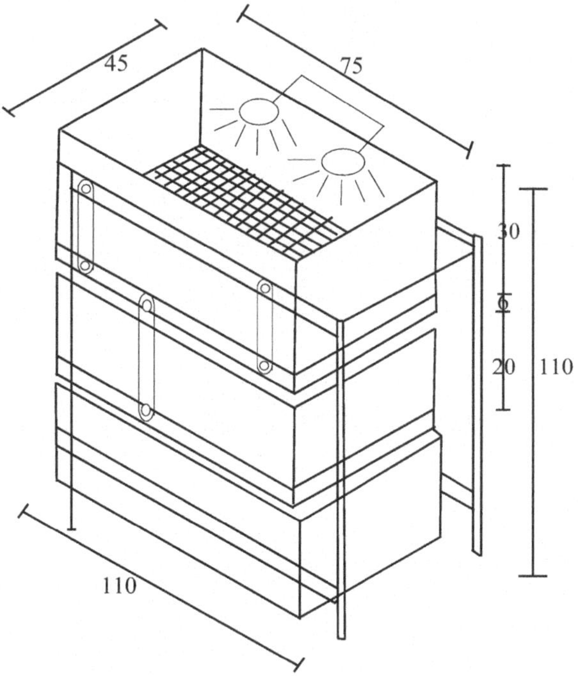

Fig. 2 Another embodiment of a VESS machine

fixed on the backside of the frame using the bearing which is perpendicular to the connecting rod. A motor is mounted on the frame which provides the movement for the pulley. The size of a typical SEVS system is 1.1 × 0.6 × 1.1 m with its first module of 0.75 × 0.45 × 0.3 m, and the second and third modules of 0.75 × 0.45 × 0.2 m. The gap between the modules is 0.06 m. As described below, the harvesting takes place in all the three modules.

The worm separation is carried out in SEVS by a source of light which drives the earthworms from the undigested substrate in the upper module through a sieve size of 6 mm to the middle module with a sieve size of 2 mm. The middle module helps in achieving the separation of the vermicast. A motor is employed to run at 200 rpm, causing vibrating action on the sieves. Due to the cam mechanism (Fig. 3), a shaking action takes place only in the first and second modules, while the third module is used as a vermicast collection unit. After the feed is added into the first module, the entire setup is kept under the light source for 5 min. The intensity of light and increase in temperature cause the earthworms to migrate toward the middle module and the undigested substrate remains in the first sieve. The harvester is made to run for 5 min so that the earthworms and the vermicast fall below to the next sieve of size 2 mm. The vermicast particles which are smaller than the pore size of the sieve fall into the collection unit.

The undigested substrate, earthworm, and vermicast from the corresponding modules are taken out by disassembling the respective containers. The earthworms and undigested substrates are then sent back to the vermireactors.

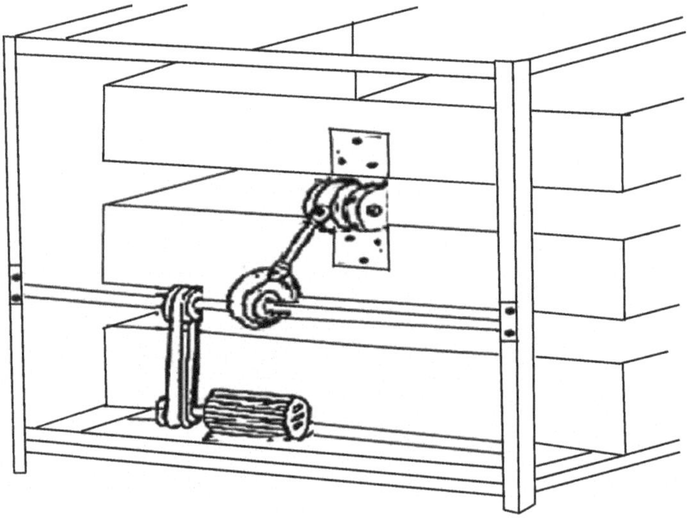

Fig. 3 The cam mechanism used in the VESS machine

3 Testing the Efficacy of the SEVS Machine

To assess the performance of the SEVS machine, it has been operated with different types of vermicast–substrate–earthworm (VES) combinations, including those resulting from the vermicomposting of paper waste, cow dung, and weeds—salvinia (*Salvinia molesta*), ipomoea (*Ipomoea carnea*), and lantana (*Lantana camara*). Before transferring these VES combinations into the upper module of the VESS, the VES were air-dried for some time and their moisture content was measured. After drying, the VES mixtures were loaded into the upper module and exposed to light for 5 min which caused the earthworms to move from the upper module to the middle one. It also aided in the separation of earthworms from the undigested substrate.

After exposing the substrate to light, the SEVS machine was made to run for 5 min. It caused the separation of the contents in their respective module. The contents of all the modules were then emptied and weighed.

For the cow dung VES, the VES was divided into different quantities of 2, 4, 6, 10, 15, and 20 kg and was air-dried for different periods: 0, 1, 4, 6, and 24 h. The substrate of each batch was then fed into the upper module of the VESS for running the machine. After the completion of the VESS operation, the contents of the upper and lower modules were emptied and weighed. The contents in the middle module were transferred back to the upper module for the second cycle of separation. When the VESS system run was over, the contents were removed from the modules and weighed.

3.1 Performance of VESS in Paper Waste Vermicast Separation

The results of duplicate runs, each with 5 kg of VES mixture, are presented in Table 1. The substrate and the earthworms were concentrated in Module 2 while the bulk of vermicast came to Module 3. The separation of most of the substrate and the

Table 1 Separation of paper-earthworm-vermicast (PEV) mixture by the SEVS machine

PEV mix	Module	Quantity (g) and proportion (%) of		
		Vermicast	Substrate	Earthworm
Run 1:5000 g	1	85 g, 6.6%	600 g, 16.2%	28, 8.0%
	2	97 g, 7.5%	3100 g, 83.6%	317, 90.1%
	3	1108 g, 85.9%	10 g, 0.3%	7, 2.0%
Run 2:5000 g	1	50 g, 2.6%	500 g, 16.3%	18, 4.5%
	2	55 g, 2.9%	2550 g, 82.9%	365, 91.9%
	3	1825 g, 94.6%	25 g, 0.8%	14, 3.5%

Table 2 Separation of salvinia-earthworm-vermicast (SEV) by the SEVS machine

SEV mix	Module	Quantity (g) and proportion (%) of		
		Vermicast	Substrate	Earthworm
Run 1 2064 g	1	7.4 g, 1.53%	490.9 g, 1.57%	3, 1.7%
	2	156.6 g, 2.5%	1018.2 g, 65.49%	166, 94.9%
	3	317.9, 66.0%	45.6 g, 2.91%	6, 3.4%
Run 2 1352 g	1	9.2 g, 2.8%	497.1 g, 39.3%	4, 2.6%
	2	77.0 g, 23.3%	743.2 g, 58.7%	144, 93.5%
	3	244.5 g, 3.9%	25.6 g, 2.0%	6, 3.9%
Run 3 1483 g	1	12.7 g, 3.6%	632.0 g, 45.5%	3, 2.7%
	2	82.4 g, 23.4%	722.9 g, 52.1%	103, 93.6%
	3	256.3 g, 3.0%	32.7 g, 2.4%	4, 3.6%
Run 4 1509 g	1	15.5 g, 3.9%	769.0 g, 54.0%	3, 2.4%
	2	67.9 g, 17.1%	622.3 g, 43.7%	118, 93.7
	3	314.3 g, 9.0%	33.8 g, 2.4%	5, 4.0%

earthworms in Module 2 is particularly useful because the mixture can be recycled in the vermireactor along with fresh feed to continue the vermicomposting.

3.2 *Performance of SEVS in the Sorting of Salvinia VES Mixture*

The results of the four replicates are summarized in Table 2. The bulk (66.0, 73.0, 73.9, and 79.0%) of the vermicast was separated in the third module, with negligible undigested substrate (<2.9%) and earthworms (<4%) coming there. As 17.1–32.5% of the vermicast was found in the second module, that mixture needed separation in another 3–5 min operation of the SEVS machine with the contents of Module 2 as input. The separation of earthworm to the second module was, however, near total (≥93.5).

3.3 *Performance of SEVS in the Sorting of the Ipomoea Vermicompost Batch*

The results of quadruplicate runs are presented in Table 3. As was seen with the salvinia VES mixture, very clean separation of earthworm (≥92.2%) occurred in the second module. The bulk of the substrate (97.3–99%) was also separated into the first or second module. As appreciable quantities (17.1–32.5%) of vermicast had remained in the second module, it called for one more run of the SEVS machine.

Table 3 Separation of ipomoea-earthworm-vermicast (IVS) by the SEVS machine

IVS mix	Module	Quantity (g) and proportion (%) of		
		Vermicast	Substrate	Earthworm
Run 1 2076 g	1	10.6 g, 1.4%	113.5 g, 8.8%	2, 1.1%
	2	341.2 g, 45.02%	1138.1 g, 88.5%	170, 96.0%
	3	406.1 g, 53.6%	34.6 g, 2.7%	5, 2.82%
Run 2 1640.4 g	1	5.7 g, 0.8%	262. 7 g, 20.2%	4, 3.1
	2	336.3 g, 49.2%	1024.9 g, 78.9%	121, 92.4%
	3	341.6 g, 50.0%	10.9 g, 0.8%	6, 4.6%
Run 3 1781.9	1	7.6 g, 1.1%	344.8 g, 25.6%	4, 3.5%
	2	425.6 g, 58.9%	989.5 g, 73.4%	106, 92.2%
	3	288.9 g, 40.0%	14.3 g, 1.1%	5, 4.3%
Run 4 1809.4 g	1	7.7 g, 0.9%	459.3 g, 31.2%	4, 2.9%
	2	329.6 g, 40.2%	995.98 g, 67.7%	128, 92.8%
	3	482.1 g, 58.8%	16.97 g, 1.2%	6, 4.3%

3.4 Efficacy of SEVS Machine in the Sorting of Lantana VES Mixture

The findings of quadruplicate runs are displayed in Table 4. Once again, clean separation ($\geq$93.3%) of earthworms occurred into the middle module, and the third module had most of the vermicast with negligible undigested substrate ($\geq$2.2%) or

Table 4 Separation of lantana-earthworm-vermicast (LVS) by the SEVS machine

LVS mix	Module	Quantity (g) and proportion (%) of		
		Vermicast	Substrate	Earthworm
Run 1 1941 g	1	8.6 g, 1.42%	146.4, 11.29%	3, 1.99%
	2	324.2 g, 53.5%	1121.7 g, 86.52%	142, 94.04%
	3	273.5 g, 45.7%	28.3 g, 2.2%	6, 3.97%
Run 2 1356 g	1	8.7 g, 2.2%	214.4 g, 17.4%	5, 2.8%
	2	115.5 g, 29.1%	1007.6 g, 81.8%	168, 93.3%
	3	273.1 g, 68.7%	10.2 g, 0.8%	7, 3.9%
Run 3 1355 g	1	10.5 g, 2.3%	219.7 g, 17.7%	3, 1.8%
	2	102.7 g, 22.5%	1012.1 g, 81.5%	161, 94.7%
	3	343.7 g, 75.2%	10.1 g, 0.8%	6, 3.5%
Run 4 1833 g	1	16.5 g, 2.5%	440.2 g, 28.1%	4, 2.2%
	2	267.4 g, 40.4%	1108.9 g, 70.9%	175, 94.1%
	3	378.3 g, 57.1%	15.4 g, 1.0%	7, 3.8%

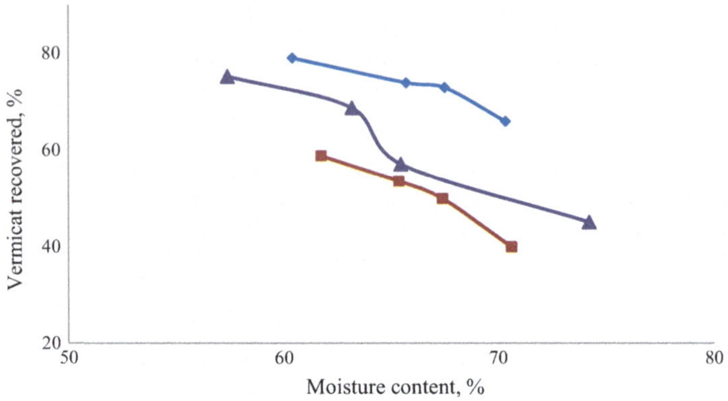

Fig. 4 Vermicast recovery as influenced by the moisture content (—■—ipomoea, —▲—lantana, —◆—salvinia)

earthworm (≤3.9%). But since the middle module had 22.5–53.5% vermicast, it called for another round of SEVS operation, with the contents of Module 2 as the input.

3.5 Effect of Moisture on the Efficiency of Vermicast Recovery From the VES Mixtures of Salvinia, Ipomoea, and Lantana

The results are displayed in Fig. 4. The extent of vermicast recovery improved by 14% in case of salvinia, 19% in case of ipomoea, and 30% in case of lantana as their moisture content was reduced by 10, 9, and 15%, respectively. It is clear that dried vermicast gets sieved more easily and completely in the SEVS machine while the wetter vermicast does not, due to its greater stickiness.

3.6 Effect of the Batch Size of the EVS Mixture and Its Drying on the Separation Caused by the SEVS Machine

Batches of cow-dung EVS mixtures of quantities ranging from 2 to 20 kg were subjected to SEVS separation with and without air-drying for periods ranging from 0 to 24 h. The results of the extent of separation of the vermicast are summarized in Table 5. In general, the level of separation improved with the duration of the drying. Batch sizes up to 15 kg gave separation reaching a maximum of 99% but in the 20 kg batches, only a maximum of 21% vermicast could be separated. The earthworm recovery (Table 6) had a similar general trend. Indeed, unlike the vermicast harvest

Table 5 Effect of VES mixture batch size and its drying on the extent of vermicast recovery by the SEVS machine

Batch size, kg	Vermicompost recovered, (%) after drying for				
	0 h	1 h	4 h	6 h	24 h
2	72	77	88	95	98
4	75	78	67	62	99
6	79	54	63	62	98
10	77	64	79	66	65
15	43	46	48	85	99
20	4	9	21	20	21

Table 6 Effect of cow dung VES mixture batch size and its drying on the extent of recovery of earthworms by the SEVS machine

Batch size	Earthworm, (%), recovered after drying the batch for				
	0 h	1 h	4 h	6 h	24 h
2	73	91	90	85	91
4	85	92	92	93	95
6	92	93	94	94	94
10	95	93	96	95	96
15	91	93	95	98	99
20	23	36	54	61	96

which had quite an erratic pattern (Table 5), earthworm recovery followed a more or less uniform pattern showing better retrieval with increasing duration of air-drying (Table 6) and even the 20 kg batch achieved 96% recovery when air-dried for 24 h.

3.7 *Effect of Subjecting the Contents of the Second Module to SEVS Separation*

The contents of the second modules of all the batches covered in Table 5 were subjected to 3 min of SEVS separation. It led to near quantitative separation in most batches as seen in Table 7. This indicates that an initial 5 min operation of SEVS, followed by 3 min of SEVS operation of the contents of the second module as obtained in the initial separation, can ensure near-complete separation of vermicast on one hand, and substrate-cum-earthworms on the other hand.

Table 7 Improvement in the vermicast recovery after the contents of the second module, obtained from the first SEVS separation, were subjected to one more SEVS separation

Size of the batch to which the second module contents belonged	Vermicompost recovery, %, corresponding to the second module contents of the batches which had been air-dried for				
	0 h	1 h	4 h	6 h	24 h
2	95	99	99	99	100
4	95	99	98	93	100
6	96	99	88	90	100
10	96	92	98	99	98
15	–	67	84	99	–
20	–	–	–	–	81

4 Summary

The paper describes the upgraded version of a substrate–earthworms–vermicompost separation (SEVS) machine recently developed by us. It is part of the technology and the associated machinery that we have introduced along with the paradigm of "high-rate vermicomposting" (HRV). The HRV enables direct and rapid vermicomposting of biowaste, especially the "phytowaste" represented by the billions of tons of biomass that is generated by terrestrial, amphibious, and aquatic weeds across the world (as elucidated in an accompanying paper in this volume). As HRV requires the frequent separation of vermicast from the unconverted substrate and earthworms, it calls for a rapid and efficient separator system. SEVS fulfills the need. The paper describes results of extensive trials on a medium-scale SEVS prototype that is capable of handling up to 20 kg at a time of vermicompost batches (mixture of vermicast, unconverted substrate, and earthworm). Vermicompost batches of five different substrates—paper waste, salvinia, ipomoea, lantana, and cow dung—were explored. The studies included the effect of batch size, moisture content of the batches, air-drying, and second stage SEVS separation on the extent of recovery of vermicast and earthworms. The results reveal the SEVS machine to be highly efficient, effective, and versatile.

Acknowledgements SAA thanks the Council of Scientific and Industrial Research (CSIR), New Delhi, for the Emeritus Scientist grant (21(1034)/16/EMR-II) which has enabled him to document the findings presented in the paper.

References

Abbasi T, Abbasi SA (2010a) Renewable energy sources: their impact on global warming and pollution. PHI Larning, New Delhi, pp x–306

Abbasi T, Abbasi SA (2010b) Production of clean energy by anaerobic digestion of phytomass—new prospects for a global warming amelioration technology. Renewable and Sustainable Energy Reviews (Elsevier), 14:1653–1659

Abbasi T, Abbasi SA (2011) Decarbonization of fossil fuels as a strategy to control global warming. Renewable and Sustainable Energy Reviews (Elsevier), 15:1828–1834

Abbasi SA, Nipaney PC (1986) Infestation by aquatic weeds of the fern genus salvinia—its status and control. Environmental Conservation, (Elsevier), 13 (3):235–241

Abbasi SA, Nayeem-Shah M, Abbasi Tasneem (2015) Vermicomposting of phytomass: limitations of the past approaches and the emerging directions. J Clean Prod 93:103–114

Abbasi T, Gajalakshmi S, Abbasi SA (2009) Towards modeling and design of vermicomposting systems: Mechanisms of composting/vermicomposting and their implications. Indian J Biotechnol 8:177–182

Abbasi T, Premalatha M, Tauseef SM, Abbasi T, Abbasi SA (2014) A Process and a system for eco-friendly and total disposal of date palm waste (DPW). Off J Pat Off 17:2412

Abbasi T, Tauseef SM, Abbasi SA (2011) The inclined parallel stack continuously operable vermireactor. Off J Pat Off 22:9571

Anuradha J, Abbasi T, Abbasi SA (2015) An eco−friendly method of synthesizing gold nanoparticles using on otherwise worthless weed *Pistia*. J Adv Res 6:711–720

Aravindan T (2014) Performance assessment of the novel vermicast earthworm substrate separator (vess) system, MTech thesis, Pondicherry University, pp 64

Banupriya G (2013) Design, fabrication and operation of a vermicast earthworm substrate separator (VESS) system on a pilot scale, MTech thesis, Pondicherry University, 79 pages

Ganaie SU, Abbasi T, Anuradha J, Abbasi SA (2014) Biomimetic synthesis of silver nanoparticles using the amphibious weed *Ipomoea* and their application in pollution control. J King Saud Univ−Sci (Elsevier), 26:222–229

Ganesh PS, Ramasamy EV, Gajalakshmi S, Abbasi SA (2005) Extraction of volatile fatty acids (VFAs) from water hyacinth using inexpensive contraptions, and the use of the VFAs as feed supplement in conventional digesters. Biochem Eng J (Elsevier) 27:17–23

Hussain, N., Abbasi, T., and Abbasi, S. A., (2015). Vermicomposting eliminates the toxicity of *Lantana* (*Lantana Camara*) and turns it into a plant friendly organic fertilizer, J Hazard Mater 298:46–57

Khan FI, Abbasi SA (1998) Techniques and methodologies for risk analysis in chemical process industries. Discovery Publishing House, New Delhi ix–364

Khan FI, Abbasi SA (1999) Assessment of risks posed by chemical industries–application of a new computer automated tool MAXCRED-III. J Loss Prev Process Ind (Elsevier), 12:455–469

Mavukkandy MO, Chakraborty S, Abbasi T, Abbasi SA (2016) A clean-green synthesis of platinum nanoparticles utilizing a pernicious weed *Lantana* (*Lantana Camara*). Am J Eng and Allied Sci 9(1):84–90

Tauseef SM, Abbasi T, Banupriya G, Banupriya D, Abbasi SA (2014) A new machine for clean and rapid separation of vermicast, earthworms and undigested substrate in vermicomposting systems. J Environ Sci Eng 56(4):495–498

Tauseef SM, Abbasi Tasneem, Banupriya D, Vaishnavi G, Abbasi SA (2013) HEVSPAR: A novel vermireactor system for treating paper waste. Off J Pat Off 24:12726

Biosorption of Safranine O Dye by SawDust

Asmita Saroha and Arpita Ghosh

Abstract Dyes used by textile industry tend to create water pollution. It not only increases COD, BOD, TDS levels, and makes water unfit to drink, but also is known to cause various health hazards (i.e., eye irritation, skin allergy, and block sunlight penetration of aquatic plants). Due to the drawbacks of conventional methods, scientists are searching for agricultural waste as suitable adsorbents for wastewater treatment. In this study, sawdust was used to remove the color of safranine o dye (C.I. 50240) from synthetic solution. Different parameters (pH, dosage, dye concentration, and particle size) were taken into consideration to study the biosorption process. The maximum color removal (approx. 98%) was quantified at pH 9 using 5 g of dosage quantity in 100 ppm of 100 ml dye solution.

Keywords Dyes · Sorption · COD · BOD

1 Introduction

Textile Industry consumes nearly 10,000 dyes annually for coloration of products like fabric, threads, etc. Dyeing process consumes only 10–25% of dye, the rest is the discharge as waste to several water bodies. Among all the dyes, azo dye is the most used one, being a member of 60–70% of all organic dyes manufactured. Dye creates havoc for aquatic plants by blocking sunlight penetration, which in turn reduces their photosynthetic activity (Hameed 2008). Dyes adversely affect pH, BOD, and COD making it unfit for drinking, irrigation, and various other household purposes. Most dyes are reported as carcinogenic, tetratogenic, and toxic in nature (Mittal and Gupta 1996). They adversely uproot normal aquatic ecosystem by altering the food web (Aksu et al. 2010).

Safranin o {color indexed as basic red 2} is a cationic dye used for commercial staining mainly by the textile and pharmaceutical industries. It has been reported to

A. Saroha · A. Ghosh (✉)
Department of Biotechnology, Manav Rachna International Institute Research and Studies, Faridabad, India
e-mail: arpi.335@gmail.com

N. A. Siddiqui et al. (eds.), *Advances in Water Pollution Monitoring and Control*, Springer Transactions in Civil and Environmental Engineering,
https://doi.org/10.1007/978-981-32-9956-6_11

cause skin allergy, respiratory irritation, and digestive tract infection (Bayazit 2014). Therefore, it is essentially important to reduce the amount of colored effluents to be dumped into various water bodies. Various conventional methods like ozonation, membrane filtration, coagulation, and flocculation were applied to clean dye generated effluent (Panswad and Wongchaisuwan 1986). The complex structure of dye made these conventional methods inefficient; moreover, they all faced disadvantages either due to their high cost of operation or because of the formation of certain types of products like aromatic amine and large sludge generation, etc.

Adsorption is one of the efficient methods for treating colored effluent. Adsorption by activated charcoal is proven to show good results after use but due to its high cost of generation, it is not economically feasible (Venkata and Sastray 1987). In an attempt to deduce an efficient but economically feasible method, agricultural biowaste was used as an adsorbent in place of cost-generating and conventional methods (Hameed 2009). Biosorption through agricultural waste is finding enormous popularity as an efficient procedure in the removal of dye and both organic and inorganic pollutants. Biosorption involves agricultural wastes that are cheap, easily available, and needs very little processing of washing, reducing the moisture content, and grinding of the adsorbent. Table 1 shows different biosorbents used for the removal of safranin o dye by researchers.

However, using different agricultural wastes as biosorbents at large-scale industrial level is still a challenge due to large-scale application and needs to be implemented.

In the present work, sawdust—a waste generated by wood and timber industry—has been used as a biosorbent. The structural composition of sawdust mainly includes cellulose, lignin, hemicelluloses, and minor amount of extraneous materials (1–5%). However, its use as a biosorbent still needs more exploration.

Table 1 Different biosorbents used for removal of safranin dye

Adsorbent	Adsorbate	Adsorption capacity	Reference
Pineapple	Safranin dye	Specific uptake: 21.7 mg/g	Mohammed et al. (2014)
Sugarcane bagasse	Safranin dye	Specific uptake: 58.853 mg/g	Farahani et al. (2015)
Alkali-treated rice husk	Safranin dye	Specific uptake: 9.77 mg/g	Chowdhury et al. (2011)
Alkali-treated mango seed	Safranin dye	Specific uptake: 31 mg/g	Malekbala et al. (2012)
Activated rice husk	Safranin dye	Removal: 82%	Gupta et al. (2006)
NaOH-treated rice husk	Safranin dye	Specific uptake: 37.97 mg/g	Chowdhury et al. (2012)

2 Materials and Methods

2.1 *Safranin o Solution*

A cationic dye safranin o (IUPAC name: 3,7-diamino-2,8-dimethyl-5-phenyl phenazinium chloride; Chemical formula: $C_{20}H_{19}ClN_4$) was used as a sample for the adsorbate (Mohammed et al. 2014). It was purchased from Thermo Electron LLS India Pvt. Ltd, Mumbai. The stock solution (1000 ppm) was prepared by diluting 1 g of safranin stain powder in 1 liter of water. Solutions of different concentrations were prepared by diluting the stock solution with distilled water in suitable volume. Natural pH of stock solution recorded around the range 8–9.

2.2 *Chemicals and Instruments*

All the fine chemicals were bought from Qualigens. To set out the range of pH, 1(N) NaOH and 1(N) HCl solutions were prepared. The pH during the experiment was analyzed by pH meter (Fisher Scientific). Experiment and process of biosorption was carried in an incubator shaker (Biosphere Corporation). RM-12C benchtop centrifuge by REMI Motors Pvt. Limited was used for separating biosorbent from aqueous dye solution after biosorption. The amount of dye removal after biosorption was analyzed via Deluxe Photo Colorimeter manufactured by Labtronics, Panchkula, India.

2.3 *Biowaste Sample*

Sawdust used as biosorbent was collected from furniture shops in the local markets around Pyali Chowk at Faridabad. The collected sawdust was rinsed with distilled water 2–3 times to remove dust and adherent particles and dried in a hot air oven at 60 °C for 48 h. It was ground in variable particle sizes (3–0.12 mm) via grinder and stored in an air-tight container to avoid moisture.

2.4 *Biosorption Procedure*

Biosorption experiments were conducted in batch mode by keeping one parameter variable and others constant. At the first set by using the same amount of biosorbent (1 g) in 100 ml dye solution (100 ppm) in different pH ranges (i.e., 3, 5, 7, and 9) were analyzed to determine the optimum pH at 30 °C temperature and 120 rpm. Similarly, experiments were conducted at a variable amount of dose concentration (3–7 g) and initial dye concentrations (10–50 ppm dye solution) keeping all the other parameters same for every set. Absorbance was analyzed of the sample dye solutions at specific time intervals (60, 120, 180, and 240 min). After experiment samples were centrifuged through benchtop centrifuge and optical density was analyzed through photocolorimeter, the amount of dye adsorbed or removed is calculated as (1)

$$\%R = 100 * \left\{ \frac{(C_0 - C_e)}{C_0} \right\} \quad (1)$$

where R is % removal of dye from aqueous solution, C_0 is the initial dye concentration, and C_e is the final concentration.

3 Results and Discussion

3.1 Adsorption Studies

Adsorption efficiency of sawdust over safranin was analyzed by varying a specific parameter (pH, temperature, biosorbent dose, dye concentration) during a particular set. A significant effect on biosorption was found by varying pH, biosorbent dose and dye concentration. However, varying temperature (range: 20–40 °C) showed no significant difference in biosorption rate. The equilibrium was found to be stabilized after 4–5 h contact time of biosorption. Slight desorption was also noticed after 5 h of biosorption in some sets.

3.2 Effect of pH

The studies were carried out on different pH ranges (i.e., 3, 5, 7, and 9) and maximum removal reported on pH 9 shows 98% of removal. The biosorption of safranin o at lower, acidic pH (Aksu et al. 2010; Bayazit 2014; Panswad and Wongchaisuwan 1986) shows slow and less amount of removal, nearly 33% in the first 2 h. As the pH increased, a more rapid and efficient removal (55%) was analyzed during the first 2 h of biosorption. The reason behind is likely the competition of H+ ion with positively charged cationic dye ions at lower pH (Ofomaja and Ho 2008). Moreover at lower pH, sawdust surface having cellulose (basic in nature) as the major constituent gets associated with H+ ions rapidly which repels the positively charged cations of safranin o dye (Farahani et al. 2015). At negative pH, OH^- ions become predominant and facilitate in adsorption by generating attraction between negatively charged adsorbent and positively charged dye. Similar trends were shown for adsorption of methyl violet on mansonia wood sawdust (Ofomaja and Ho 2008) (Figs. 1 and 2).

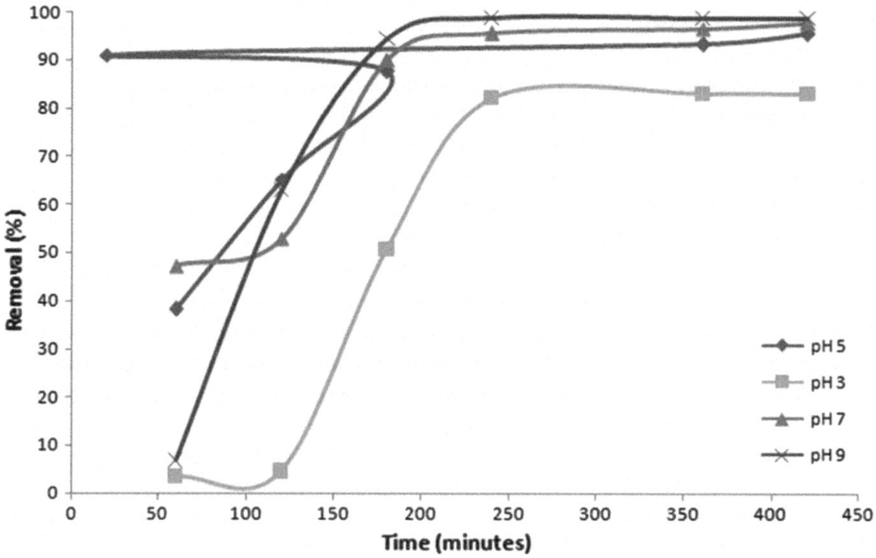

Fig. 1 Showing removal at different pH

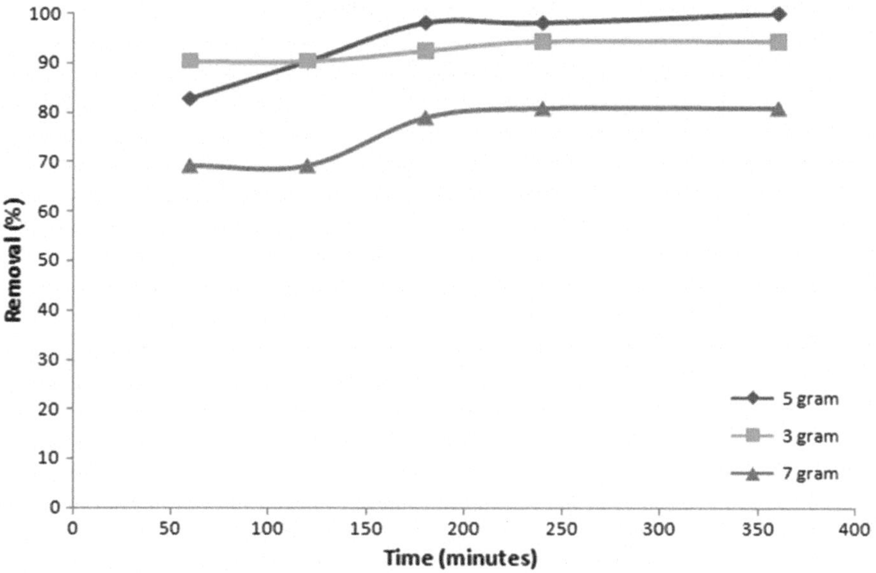

Fig. 2 Showing removal at varying biosorbent dosages

Effect of Dose Amount

Three Variants of dosage amount were considered as (3, 5, and 7 g). All three of them reported desorption in the first 3 hours of adsorption but after that, a stable adsorption of 98% was reported. On increased amount of biosorbent, initially, there was a decline in the adsorption rates but after 4 h, the maximum removal (98%) was quantified in the set having more biosorbent. However, there is no significant difference in adsorption percentage found between 5 and 7 g.

Effect of Dye Concentration

Effect of varying dye solutions (10 ppm, 25 ppm, and 50 ppm) was analyzed showing maximum absorption of 98%, 97%, and 95%, respectively. A decreasing trend of biosorption was reported on increasing dye concentration. The reason possibly is that the saturation at the biosorbent sites makes them inefficient to adsorb more dye particles (Fig. 3).

Effect of Temperature

The effect of temperature was analyzed at 20–40 °C. A slight increase in the removal of dye from 72 to 74% was analyzed at a higher temperature. However, the difference was not significant even though it demarks the exothermic nature of adsorption. The same pattern is noticed in the removal of basic dye by tree fern (Ho et al. 2005) (Fig. 4).

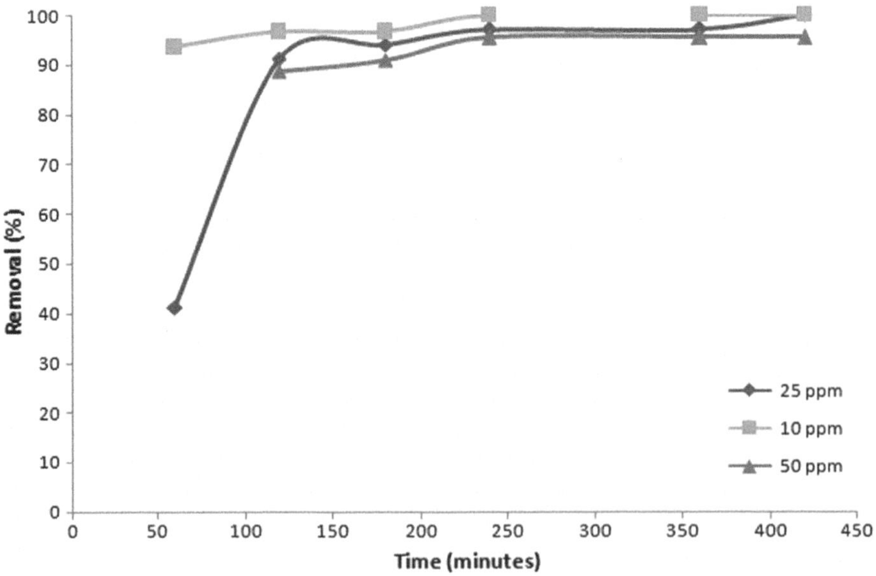

Fig. 3 Depicting removal on increasing dye concentration

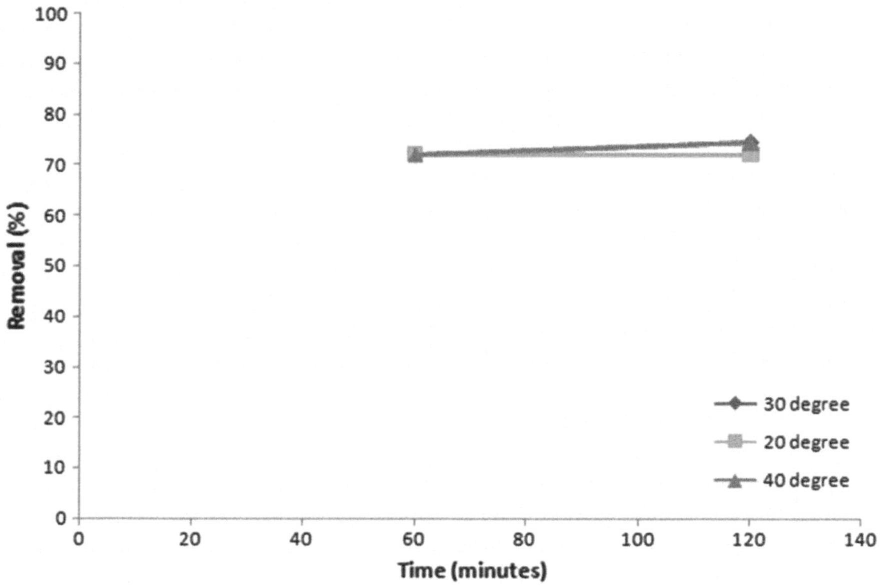

Fig. 4 Showing effect of temperature

Desorption and Reuse of Biosorbent

Desorption of used biosorbent is done by keeping them in HCl and NaOH solution overnight and then let it dry at room temperature for 2 to 3 days after which it was reused and its reuse efficiencies were analyzed. The maximum adsorption found was 57% in case of NaOH and nearly 25% in the case of HCl. The probable reason could be an addition in the basic nature of cellulose adsorbent making it more efficient for biosorption (Ghosh et al. 2017; Ghosh and Das 2014) (Fig. 5).

4 Conclusion

Sawdust was found as an efficient and inexpensive biosorbent showing 98% removal from 100 ppm dye solution. Biosorption was found to be more efficient at high pH, till 5 g of biosorbent dose and at lower initial dye concentration. Higher temperature was also found more favorable for adsorption. A high reuse efficiency of 57% was also analyzed making it an efficient biosorbent. The above work concludes that sawdust can be used for treating textile wastewater, which also adds sustainability to waste management.

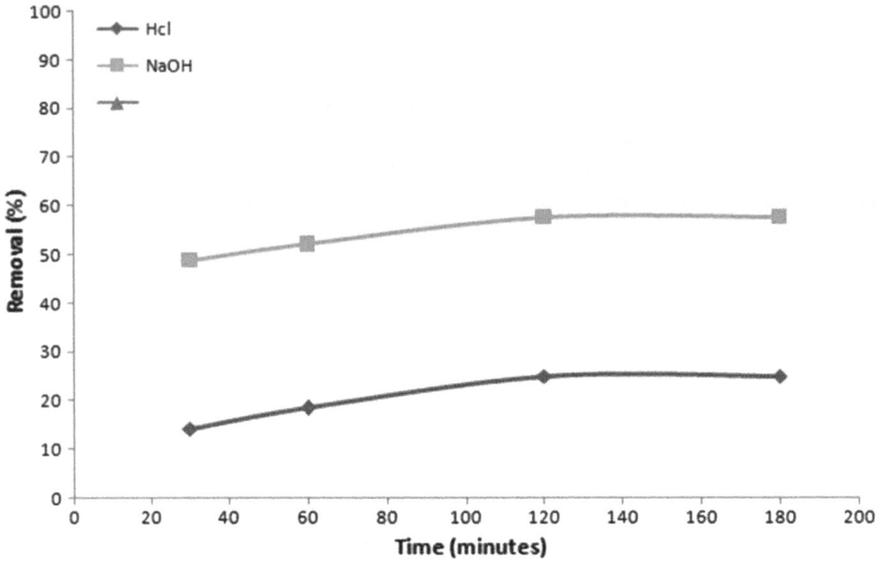

Fig. 5 Depicting reuse efficiencies of biosorbent

References

Aksu Z, Ertuğrul S, Dönmez G (2010) Methylene Blue biosorption by Rhizopus arrhizus: Effect of SDS (sodium dodecylsulfate) surfactant on biosorption properties. Chem Eng J 158(3):474–481

Bayazit ŞS (2014) Investigation of Safranin O adsorption on superparamagnetic iron oxide nanoparticles (SPION) and multi-wall carbon nanotube/SPION composites. Desalination Water Treat 52(37–39):6966–6975

Chowdhury S, Misra R, Kushwaha P, Das P (2011) Optimum sorption isotherm by linear and nonlinear methods for safranin onto alkali-treated rice husk. Bioremediat J 15(2):77–89

Chowdhury S, Mishra R, Kushwaha P, Saha P (2012) Removal of safranin from aqueous solutions by NaOH-treated rice husk: thermodynamics, kinetics and isosteric heat of adsorption. Asia-Pac J Chem Eng 7(2):236–249

Farahani M, Kashisaz M, Abdullah SRS (2015) Adsorption of safranin O from aqueous phase using sugarcane bagasse. Int J Ecol Sci Environ Eng 2:17–29

Ghosh A, Das P (2014) Optimization of copper adsorption by soil of polluted wasteland using response surface methodology. Indian Chem Eng 56(1):29–42

Ghosh A, Dastidar MG, Sreekrishnan TR (2017) Bioremediation of chromium complex dyes and treatment of sludge generated during the process. Int Biodeterior Biodegradation 119:448–460

Gupta VK, Mittal A, Jain R, Mathur M, Sikarwar S (2006) Adsorption of Safranin-T from wastewater using waste materials—activated carbon and activated rice husks. J Colloid Interface Sci 303(1):80–86

Hameed 22B (2008) Equilibrium and kinetic studies of methyl violet sorption by agricultural waste. J Hazard Mater 154(1–3):204–212

Hameed BH (2009) Grass waste: a novel sorbent for the removal of basic dye from aqueous solution. J Hazard Mater 166(1):233–238

Ho YS, Chiang TH, Hsueh YM (2005) Removal of basic dye from aqueous solution using tree fern as a biosorbent. Process Biochem 40(1):119–124

Malekbala MR, Soltani SM, Yazdi SK, Hosseini S (2012) Equilibrium and kinetic studies of safranine adsorption on alkali-treated mango seed integuments. Int J Chem Eng Appl 3(3):160
Mittal AK, Gupta SK (1996) Biosorption of cationic dyes by dead macro fungus Fomitopsis carnea: batch studies. Water Sci Technol 34(10):81–87
Mohammed MA, Ibrahim A, Shitu A (2014) Batch removal of hazardous safranin-O in wastewater using pineapple peels as an agricultural waste based adsorbent. Int J Environ Monit Anal 2(3):128–133
Ofomaja AE, Ho YS (2008) Effect of temperatures and pH on methyl violet biosorption by Mansonia wood sawdust. Biores Technol 99(13):5411–5417
Panswad T, Wongchaisuwan S (1986) Mechanisms of dye wastewater colour removal by magnesium carbonate-hydrated basic. Water Sci Technol 18(3):139–144
Venkata RB, Sastray CA (1987) Removal of dyes from aqueous solutions by cellulosic waste orange peel. Indian J Environ Prot 7:363–376

Recent Developments in Copper and Iron Based Dyes as Light Harvesters

Radha Mishra

Abstract Application of transition metal complexes as light harvesters or sensitizers for solar cells is an active research field. Second- and third-row transition metal based sensitizers such as Ru, Ir and Pt have been extensively investigated and Ru-polypyridyl dyes have been reported to be the most efficient. But these metals are very expensive and scarce. This has intensified the investigations for alternative sensitising dyes based on more abundant first-row transition metals such as copper, iron and nickel. However, these metal complexes have excited states with a very short lifetime unlike their second and third row analogues due to which electron injection in the TiO_2 layer becomes a limiting factor resulting in lower efficiency. To develop efficient DSSCs, the first-row transition metal sensitizer molecule design should focus simultaneously on optimising their chemical properties to enhance spectral coverage and photo-physical properties to enhance the lifespan of excited states. Recently, theoretical studies of electronic and spectral properties of sensitising dye molecules using Density functional theory have served as an important guide to design sensitizers with improved efficiency. A few complexes of Cu(I) and Fe(II) N-heterocylic carbenes have been reported to have great potential as efficient sensitizers. This paper summarises various strategies adopted by researchers and the efforts devoted to exploring the possibility of using these metal complexes as light harvesters in recent years.

Keywords Sensitising dyes · DFT · Charge transfer states

1 Introduction

High-efficiency dye sensitised solar cells popularly known as Gratzel cells (O'Regan 1991) were the first example of an economically competitive alternative to silicon-based solar cells. A DSSC is a photo electrochemical device which depends on the sensitisation of a wide-band semiconductor by a dye sensitizer to generate electricity.

R. Mishra (✉)
Chemistry Department, Meerut College, Meerut, India
e-mail: radhamishra2001@gmail.com

N. A. Siddiqui et al. (eds.), *Advances in Water Pollution Monitoring and Control*,
Springer Transactions in Civil and Environmental Engineering,
https://doi.org/10.1007/978-981-32-9956-6_12

The sensitizer is one of its critical components that absorbs light energy and injects the photo-excited electron into the conduction band of the semiconductor (Zhang et al. 2014). The electronic structure of the sensitizer molecule, nature of anchoring and ancillary group and its HOMO and LUMO energy levels affect the light absorbance and electron injection, and hence the composite performance of the solar cell. As such development of metal complex dyes for use as sensitizers is an active field of research. Ruthenium polypyridyl complex dyes have been investigated extensively and dyes like N3 dye (Chiba et al. 2006; Gratzel 2004) and N719 (Nazeeruddin et al. 1993) have been reported to have reached 11% efficiency. However, their large-scale application is restricted due to high cost, limited availability and toxicity of ruthenium metal. This has motivated researchers to explore the possibility of using complexes of the first-row transition metals as inexpensive, abundant and non-toxic sensitizers in DSSCs.

2 Design Criteria for a Dye Sensitizer Molecule

An efficient dye sensitizer should have high absorbance in the visible region. Its HOMO–LUMO levels and redox potentials should appropriately match with the energy levels of the semiconductor and electrolyte for electron injection and regeneration respectively and should be able to sustain numerous redox turnovers under light irradiation (Gratzel 2001). Further, the sensitizer should be photo stable and chemically bound to the semiconductor. The MLCT transitions observed in these dye sensitizers play an important role in charge injection to TiO_2 surface. Despite first-row transition metals fulfilling most of the above criteria and not presenting abundance or toxicity problems, research into first-row transition metal sensitizers has been comparatively limited. This is due to the fact that their excited states have an extremely short lifetime, probably due to crossing from charge-transfer state to closely spaced d-d excited states followed by non-radiative thermal decay (Linfoot et al. 2011). This factor limits electron injection into TiO_2 leading to low efficiency.

The sensitizer design strategies implemented to enhance sensitising efficiency should, therefore, focus on extending the lifetime of the charge transfer states most probably by preventing crossing to d-d excited states. This has been reported to be achieved by either having complexes with completely filled d shells or by having a strong crystal field so as to raise the energy of d-d excited states. Based on these two strategies, several copper(I), Iron(II) and nickel(II)-based dyes have been recently investigated as sensitizers with great potential (Bozic-Weber et al. 2011). This review highlights the various strategies adopted by researchers in recent years in molecular designing of sensitizers based on these first-row transition metals so as to develop efficient, cost-effective and environment-friendly DSSCs.

3 Cu Sensitizers

Cu(I) bis (diimine) complexes have been reported to possess photo-physical properties similar to that of Ru(II) diimine complexes (Armaroli 2001; Armaroli et al. 2007). The closed-shell d^{10} Cu(I) dyes have MLCT excited state with considerable lifetime (Armaroli et al. 2006, 2007) resulting due to the excitation of a d electron from metal bonding orbital to ligand π^* antibonding orbital. However, the MLCT excited state in ruthenium(II) and copper(I) sensitizers has rather different characteristics. In ruthenium compounds, the equilibrium geometry between the ground and excited states do not show much variation. However, in Cu(I) complexes the conformational geometry changes from distorted tetrahedral in the ground state to tetragonally flattened geometry in an excited state (Robertson 2008). Such geometric distortions slow down the electron injection into the semiconductor and allow a non-radiative relaxation of the excited state (Armaroli 2001; Armaroli et al. 2007). Due to this reason, the early investigations on homoleptic Cu(I)-based dye for DSSC by Sauvage and his co-workers (Alonso-Vante et al. 1994) in 1994 were reported to have much lower efficiency.

Investigations on a number of phenanthroline- and bipyridine-based copper(I) complexes have led researchers to the conclusion that efficient sensitizers can be developed by structural modifications of ligands to tune their photo-physical and electrochemical properties. In 2002, Sakaki and co-workers (Sakaki et al. 2002) reported a new Cu(I) dye with 5,5' carboxy substituted bipyridyl ligands to ensure rapid electron injection with a conversion efficiency of 2.5%. This led to the conclusion that Cu(I) dyes have great potential as sensitizers for solar cells and provided evidence that molecular designing through structural changes in ligand and optimisation could improve efficiency. Despite these reports, little progress was made in Cu-based DSSCs and this remained an inactive field until 2008 when Constable, Housecroft and co-workers (Bessho et al. 2008) synthesised Cu(I) sensitizers with 6,6'-dimethyl 2,2'-bipyridyl ligands and found these dyes to be effective sensitizers for TiO_2. They emphasised that the substituents at the 6 and 6' positions in 2,2' bipyridyl ligands are a basic structural requirement to enhance the stability of Cu(I) dyes. The stearic bulk of 6,6' substituents effectively stabilises Cu(I) with respect to oxidation to Cu(II); thus avoids flattening distortion and produces long-lived excited states. Subsequently, many homoleptic Cu(I) bipyridyl dyes with various substituents (Constable et al. 2009; Bozic-Weber et al. 2011; Yuan et al. 2012; Colombo et al. 2013; Wills et al. 2013) were investigated and the one reported to have the highest performance up to now is the dye with 6,6'dimethyl-2,2'bipyridine-4,4'-dibenzoic acid ligands (Yuan et al. 2012). Lu et al. (2010) and Lopez et al. (2010) theoretically studied several homoleptic Cu(I)-based dyes by DFT and TDDFT approach to describe their geometries and spectral properties. They emphasised the importance of theoretical calculations in the design of new copper sensitizers with improved performances.

In homoleptic systems, the directional movement of photo-generated electrons across the dye is hampered due to high level of symmetry. This led to an emphasis on using heteroleptic Cu(I) dye sensitizers having one ligand with a group suitable for anchoring the TiO_2 surface and the second ligand tuned structurally for light absorption. Thus, there is a push–pull system that enables the charge separation and provides directionality for electron injection to the TiO_2 conduction band. In 2010, C. L. Linfoot et al. reported the first heteroleptic Cu(I) sensitising dye [Cu(POP)dcdpy][BF_4] but with very low efficiency (Linfoot et al. 2010). The subsequent researches focused on the methodology of developing heteroleptic Cu(I) dyes and to synthesise D-π-A type sensitizers. Constable and Housecroft (2015) group developed a 'Surfaces-as ligands, surfaces-as complexes' strategy while Odobel et al. (Sandroni et al. 2013a, b) used 'HETPHEN concept' to develop stable heteroleptic Cu(I) complexes. Odobel et al. applied computational techniques such as DFT and TDDFT to elucidate the electronic structure and spectral properties of newly designed and synthesised Cu(I)-based sensitizers. In the past few years, effects of different ancillary groups (Hewat et al. 2014) and anchoring groups (Hewat et al. 2014), introduction of various substituents (Malzner et al. 2014; Brunner et al. 2015), introduction of co-adsorbents (Brauchli et al. 2014) and of incorporating hole transport motifs (Bozic–Weber et al. 2013b) on the efficiency of copper(I) DSSCs have been intensively investigated and DFT and TDDFT have been increasingly used to support the results. Detailed reviews by Housecroft (2015) and Magani (2016) provide a wealth of information on the strategy and design of heteroleptic Cu(I) sensitizers. These investigations have led to the conclusion that the development of efficient sensitizers is possible by accelerating electron injection either by using electron-releasing substituents on the ancillary ligand or by using a more rigid anchoring ligand to extend the lifetime of the excited state (Sandroni et al. 2013b).

4 Iron Sensitizers

Fe(II) polypyridyl complexes such as $[Fe(dcbpy)_2(CN)_2]$ as sensitizers were first investigated by Ferrere and Gregg (1998) in 1998. Later, Meyer (Yang et al. 2002) and co-workers studied $Na_2[Fe(bpy)(CN)_4]$ complexes. They observed two types of charge transfer mechanisms—(i) Direct mechanism in which an electron from the sensitising dye is directly injected in conduction band of TiO_2 and (ii) Indirect mechanism involving interfacial electron transfer (IET) from initially excited charge-transfer state of dye into the TiO_2 (Ferrere and Gregg 1998; Yang et al. 2002). Ghosh et al. (1998) observed a direct mechanism of charge transfer in $[Fe(CN)_6]^{4-}$. Iron complexes are considered to be a possible alternative to ruthenium sensitizers as it belongs to the same group as ruthenium. However, iron complexes unlike their ruthenium counterparts sensitise TiO_2 very inefficiently inspite of an intense MLCT absorption. This is due to the fact that the photoactive MLCT state is extremely short-lived and gets deactivated into photo inactive metal centred (MC) states due to intersystem crossing (ISC) in sub-picoseconds (Monat and McCusker 2000; Juban et al. 2006; Smeigh et al. 2008). In short, the ultrafast ISC phenomenon triumphs over the relatively slow interfacial electron transfer leading to inefficient sensitization.

Hence, the development of Fe(II) complexes capable of IET at a faster rate as compared to ISC is an important aspect of improving sensitising efficiency. An understanding of different structural aspects that influence the rates of the ISC and IET, therefore, is the key to the development of more efficient iron-based sensitising dyes. Computational methods such as DFT and TDDFT are extremely helpful for such investigations because they enable evaluation of the effect of systematic structural modifications on light harvesting and other photo-physical properties. The efficiency of Fe(II) sensitizers can be improved by adopting basically two strategies, both of which have been explored in recent years by different research groups with considerable success. The first approach focuses on making IET faster than ISC process by increasing the rate of IET through structural modification in the ligands of sensitising dyes (Bowman et al. 2013, 2015a; Jakubikova and Bowman 2015; Nance et al. 2015). Bowman et al. (2013) studied three different dye-TiO_2 systems having polypyridine scaffolds functionalised with three different electron-donating groups and concluded that electron-donating substituents may improve electron injection and hence the efficiency of Fe(II) polypyridine-based dyes (Bowman et al. 2013). Later, the same research group computationally investigated the effect of various linker groups in $[Fe(bpy\text{-}L)_2(CN)_2]$ dye and determined the rate of IET in each case (Bowman et al. 2013). The molecular simulations predicted that the hydroxamate linker instead of carboxylic acid linker will have a high rate of IET and hence Fe(II) polypyridines with hydroxamate will be more efficient sensitising dyes.

The second approach towards enhancing the efficiency of these sensitising dyes focuses on slowing down the ISC process and thus increasing ^{3} MLCT lifetime. This was reported to be achieved by modifying the ligands in Fe(II) polypyridines to increase the ligand field strength. These modifications included changing the donor atom in ligands (Dixon et al. 2015; Mukherjee et al. 2015) or altering the ligand framework geometry (Mengel et al. 2015; Liu et al. 2015). Bowman et al. systematically modified donor atom and ligand framework to compare the relative ligand field strengths in various iron (II) dyes and concluded that formation of Fe–C bonds leads to the strongest ligand field. Liu and co-workers studied $[Fe(CNC)_2]^{2+}$ dye [CNC = 2,6-bis(imidazol-2-ylidene)pyridine] with strongly σ donating N-heterocyclic carbene ligands with an aim to raise the, e_g orbital energy to slow down ISC. This study reported a record ^{3}MLCT lifetime of t = 9 picoseconds (Liu et al. 2013), which was improved further by the same group to t = 26 picoseconds recently (Liu et al. 2016). Thus extending the lifetime of ^{3}MLCT state in iron complexes is an important strategy for improving efficiency of Fe(II) sensitizers.

5 Nickel Sensitizers

The literature on the use of nickel as sensitizers for DSSCs is scanty as some of the initially investigated Ni complex dyes reported insignificant photovoltaic activity. K. Neuthe used heteroleptic Ni(II) diimine dithiolenes as co-sensitizers (Neuthe et al. 2014) to known visible light-harvesting dyes for extending the spectral window to infra-red region. In 2011, Linfoot et al. reported [Ni(dcdpy)(qdt)] and [Ni(decbpy)(qdt)] as possible sensitizers and studied their electronic properties using DFT and TDDFT (Bozic–Weber et al. 2013a). These complexes exhibit intense visible transition between 500 and 600 nm that can be assigned to charge transfer from qdt to diimine ligand. However, the conversion efficiency is observed to be quite low in spite of satisfactory light harvesting probably due to short-lived excited state. This suggests that highly efficient Ni-based sensitizers can be designed if excited state lifetime is enhanced by suitable molecular design.

6 Conclusion

The research in first-row transition metal dye sensitizers for solar cell applications has tremendously progressed over the past decade. This review systematically summarises the innovative strategies employed by various groups for enhancing the efficiency of the dye sensitizers based on more available, cost-effective and non-toxic metals. Some complexes of Cu(I) and Fe(II) have led to promising results. However, the device performances with these sensitizers are still less than those of Ru(II) sensitizers. The development of first-row metal complexes as efficient and inexpensive sensitizers at commercial scale in the future requires important challenges to be tackled. Optimising the chemical properties to broaden the spectral coverage and enhancing photo-physical properties to enhance excited state lifetime and increasing the stability are some research areas that need continuous efforts.

References

Alonso-Vante N, Nierengarten J-F, Sauvage J-P (1994) J Chem Soc Dalton Trans 11:1649–1654
Armaroli N (2001) Chem Soc Rev 30:113–124
Armaroli N, Accorsi G, Holler M, Moudam O, Nierengarten J-F, Zhou Z, Wegh RT, Welter R (2006) Adv Mater 18:1313–1316
Armaroli N, Accorsi G, Cardinali F, Listorti A (2007) Top Curr Chem 280:69–115
Baldenebro-lopez J, et al (2014) Polyhedron 82:33–36
Bessho T, Constable EC, Gratzel M, Hernandez Redondo A, Housecroft CE, Kylberg W, Nazeeruddin MK, Neuburger M, Schaffner S (2008) Chem Commun 3717–3719
Bowman DN, Blew JH, Tsuchiya T, Jakubikova E (2013) Inorg Chem 52:8621–8628
Bowman DN, Mukherjee S, Barnes LJ, Jakubikova E (2015a) J Phys: Condens Matter 27:134205
Bowman DN, Bondarev A, Mukherjee S, Jakubikova E (2015b) Inorg Chem 54:8786–8793

Bozic-Weber B, Constable EC, Housecroft CE, Kopecky P, Neuburger M, Zampese JA (2011) Dalton Trans 40:12584
Bozic-Weber B, Constable EC, Housecroft CE (2013a) Coord Chem Rev 257:3089–3106
Bozic-Weber B, Brauchli SY, Constable EC, Fuerer SO, Housecroft CE, Wright IA (2013b) Phys Chem Chem Phys 15:4500–4504
Brauchli SY, Malzner FJ, Constable EC, Housecroft CE (2014) RSC Adv 4:62728–62736
Brunner F, Klein YM, Keller S, Morris CD, Prescimone A, Constable EC, Housecroft CE (2015) RSC Adv 5:58694–58703
Chiba Y, Islam A, Watanabe Y, Komiya R, Koide N (2006) Jpn J Appl Phys 45:L638–L640
Colombo A, Dragonetti C, Roberto D, Valore A, Biagini P, Melchiorre F (2013) Inorg Chim Acta 407:204–209
Constable EC, Redondo AH, Housecroft CE, Neuburger M, Schaffner S (2009) Dalton Trans 6634–6644
Dixon IM, Alary F, Boggio-Pasqua M, Heully JL (2015) Dalton Trans 44:13498–13503
Ferrere S, Gregg BA (1998) J Am Chem Soc 120:843
Ghosh HN, Asbury JB, Weng Y, Lian TJ (1998) Phys Chem B 102:10208
Gratzel M (2001) Nature 414:338–344
Gratzel M (2004) J Photochem Photobiol, A 168:235–237
Hewat TE, Yellowlees LJ, Robertson N (2014) Dalton Trans 43:4127–4136
Housecroft CE, Constable EC (2015) Chem Soc Rev 44:8386–8398
Jakubikova E, Bowman DN (2015) Acc Chem Res 48:1441–1449
Juban EA, Smeigh AL, Monat JE, McCusker JK (2006) Coord Chem Rev 250:1783
Linfoot CL, Richardson P, Hewat TB, Moudam O, Forde MM, Collins A, White F, Robertson N (2010) Dalton Trans 39:8945–8956
Linfoot CL, Richardson P, McCall KL, Durrant JR, Morandeira A, Robertson N (2011) Sol Energy 85:1195–1203
Liu Y, Harlang T, Canton S, Chabera S, Suarez Alcantara K, Fleckhaus A, Goransson E, Corani E, Lomoth A, Sundstrom R, Warnmark VK (2013) Chem Comm 6412
Liu Y, Kjer KS, Fredin LA, Chabera P, Harlang T, Canton SE, Lidin S, Zhang J, Lomoth R, Bergquist KE, Persson P, Wärnmark K, Sundström V (2015) Chem Eur J 21:3628–3639
Liu L, Duchanois T, Etienne T, Monari A, Beley M, Assfeld X, Haacke S, Philippe C (2016) Gros, Phys Chem Chem Phys 18:12550–12556
Lopez JB, et al (2012) Int J Mol Sci 13:16005–16019
Lu X, et al (2010) J Phys Chem A 114:1178–1184
Magni M, Biagini P, Colombo A, Dragonetti C, Roberto D, Valore A (2016) Coord Chem Rev 322:69–93
Malzner FJ, Brauchli SY, Constable EC, Housecroft CE, Neuburgur M (2014) RSC Adv 4:48712–48723
Mengel AKC, Förster C, Breivogel A, Mack K, Ochsmann JR, Laquai F, Ksenofontov V, Heinze K (2015) Chem Eur J 21:704–714
Monat JE, McCusker JK (2000) J Am Chem Soc 122:4092
Mukherjee S, Bowman DN, Jakubikova E (2015) Inorg Chem 54:560–569
Nance J, Bowman DN, Mukherjee S, Kelley CT, Jakubikova E (2015) Inorg Chem 54:11259–11268
Nazeeruddin MK, Kay A, Rodicio I, Humphry-Baker R, Muller E, Liska P, Vlachopoulos N, Gratzel M (1993) J Am Chem Soc 115:6382–6390
Neuthe K, Popeney CS, Bialecka K, Hinsch A, Sokolowski A, Veurmann W, Haag R (2014) Polyhedron 81:583–587
O'Regan M (1991) Gratzel. Nature 353:737–740
Robertson N (2008) Chem Sus Chem 1:977
Sakaki S, Kuroki T, Hamada T (2002) J Chem Soc Dalton Trans 840–842
Sandroni M, Kayanuma M, Rebarz M, Akdas-Kilig H, Pellegrin Y, Blart E, Le Bozec H, Daniel C, Odobel F (2013a) Dalton Trans 42:14628–14638

Sandroni M, Kayanuma M, Rebarz M, Akdas-Kilig H, Pellegrin Y, Blart E, Le Bozec H, Daniel C, Odobel F (2013b) Dalton Trans 42:10818–10827
Smeigh AL, Creelman M, Mathies RA, McCusker JK (2008) J Am Chem Soc 130:14105
Wills KA, Mandujano-Ramirez HJ, Merino G, Mattia D, Hewat T, Robertson N, Oskam G, Jones MD, Lewis SE, Cameron PJ (2013) RSC Adv 3:23361–23366
Yang M, Thompson DW, Meyer GJ (2002) Inorg Chem 41:1254–1262
Yuan Y-J, Yu Z-T, Zhang J-Y, Zou Z-G (2012) Dalton Trans 41:9594
Zhang W, Jin X, Zhou J, Tang G, Peng D, Hu J, Zhong C (2014) J Organomet Chem 749:26–33

Green Synthesis of Silver Nanoparticles Using *Cucurbita pepo* Leaves Extract and Its Antimicrobial and Antioxidant Activities

Deepak Kumar, Shefali Arora, Abhishek Kumar, Abdullah, Mohd. Danish and Kanchan Deoli Bahukhandi

Abstract The present study is aimed on the plant-based synthesis of silver nanoparticles by $AgNO_3$ and by use of the phytoconstituents of ethanol extract from the leaves of *Cucurbita pepo*. The silver nanoparticles of average size in the range of 5–100 nm have been synthesized by easy, fast and environmentally friendly method. Silver nanocrystals were characterized using SEM analysis and UV Spectrophotometer. The silver nanoparticles and ethanol extract undergo antimicrobial and antioxidant activities. The results showed that silver nanoparticles showed significant antimicrobial and mild antioxidant activities, respectively.

Keywords *Cucurbita pepo* · Antimicrobial · Antioxidant · Silver nanoparticles · DPPH

1 Introduction

The plant extracts was used for the treatment of many diseases by medicinal practitioners since oldest time. These plants based products are environmentally friendly, safe, non-toxic, cost-effective than synthetic drugs (D'britto et al. 2012).

The plant materials based synthesis of silver nanoparticles (AgNPs) have lots of advantages for pharmaceutical and biomedical applications because of the non-toxic synthesis protocols. Chemical synthesis of AgNPs uses toxic chemicals which showed adverse effects as such chemicals adsorbed on the surface of nanoparticles.

D. Kumar (✉) · A. Kumar · Abdullah · Mohd. Danish
Department of Pharmaceutical Chemistry, Dolphin PG Institute of Biomedical and Natural Sciences, Dehradun, Uttarakhand, India
e-mail: deepsingh2304@gmail.com

S. Arora
Department of Chemistry, University of Petroleum and Energy Studies, Dehradun, Uttarakhand, India

K. D. Bahukhandi
Department of Health Safety and Environment, University of Petroleum and Energy Studies, Dehradun, Uttarakhand, India

N. A. Siddiqui et al. (eds.), *Advances in Water Pollution Monitoring and Control*, Springer Transactions in Civil and Environmental Engineering,
https://doi.org/10.1007/978-981-32-9956-6_13

Plant-based herbal synthesis of nanoparticles are safe, non-toxic, eco-friendly and provides more benefits over chemical and physical methods (Goodsell 2004). The advancement of plant-based synthesis of nanoparticles is an emerging a branch of nanotechnology because of many applications in the fields of Science especially medicine, physics and chemistry (Song and Kim 2008). AgNPs have been used in many areas like sensors, textiles, detectors, nanocomposites, catalysis, agriculture, waste water treatment and more important in medicine (Shankar et al. 2003a, b; Hettiarachchi and Wickramarachchi 2011). The synthesis of AgNPs is speedy, cost saving, eco-friendly and single-step synthesis procedure (Huang et al. 2007).

Cucurbita pepo belongs to the genus Cucurbita in Cucurbitaceae family. The local name of *Cucurbita pepo* is Saraiki in Hindi, Squash in English and Kadoo in Urdu. *Cucurbita pepo* broadly occurs throughout tropical and subtropical areas of the globe, especially America, Africa, Asia, and Australia.

Cucurbita pepo is traditionally used in many countries to treat several diseases like anti-inflammatory, antiviral, analgesic urinary disorders, antiulcer, antidiabetic and antioxidant (Martha and Gutierrez 2016). Ayurveda and Chinese systems have used different parts of the plant as traditional medicine. *Cucurbita pepo* is one among the medicinal plant which contains different phytoconstituents that gives health benefits (Caili and Quanhong LI 2006). The phytoconstituents reported in seeds of *Cucurbita pepo* are cucurbitacin, cardiac glycosides, terpenoids and resins (Rahman 2013; Nawirska-Olszanska et al. 2013). Flower contains phenolic compounds such as quercetin, 4-dihydroxy methyl benzoate, Isorhamnetin, myricetine 3, 6, 3, trimethoxyflavone (Mohamed et al. 2009). Fruit contains fatty acids, linoleic acid, calotropoleanly ester and cholesterol (Badr et al. 2011). The present study was investigated for the synthesis of AgNPs using the ethanol extract of leaves of *Cucurbita pepo* and its antimicrobial and antioxidant activities.

2 Materials and Methods

2.1 Collection and Identification of Plant Materials

Leaves of *Cucurbita pepo* were collected from Bandahari village of Saharanpur, UP, India. Plant material was deposited and authenticated by Dr. Sunita Garg, Chief Scientist, (NISCAIR-CSIR), New Delhi, India with reference number NISCAIR/RHMD/Consult/2015/2819/12.

2.2 Preparation of Ethanol Extract of Leaves of Cucurbita pepo

The leaves were washed with water to remove other undesirable substance and dried under shade. The air-dried leaves (100 gm) of *Cucurbita pepo* were crushed and extracted with ethanol. The extract was evaporated till dryness and concentrated.

2.3 Test of Phytoconstituents

Ethanol extract was subjected for qualitative phytochemical tests for presence of carbohydrate, alkaloid, protein and amino acids, sterols, phenolic compounds and saponin (Tiwari et al. 2011).

2.4 Synthesis of AgNPs from Ethanol Extract

The ethanol extract of *Cucurbita pepo* was subjected to the synthesis of AgNPs. 1 mM aqueous solution of silver nitrate ($AgNO_3$) was prepared. 300 mg of ethanol extract was dissolved with little amount of ethanol and volume makeup to 100 ml by water. To 60 ml of *Cucurbita pepo* ethanol extract solution, 10 ml of silver nitrate solution was added, mixed and kept at room temperature to react (Parashar et al. 2009).

2.5 Characterization of AgNPs

The bioreduction of $AgNO_3$ was confirmed by scanned UV-Vis spectra of reaction mixture sample of absorption maximum at the wavelength of 200–700 nm in Thermo Scientific UV spectrophotometer. Further, the bioreduced reaction mixture was analysed by SEM studies by SEM (JEOL, Model JFC-1600).

2.6 Antimicrobial Activity of Ethanol Extract and AgNPs

Ethanol extract and AgNPs were tested for antibacterial and antifungal activities at a concentration of 200 mg/ml (Barry and Fibiger 1976).

2.7 Antibacterial Activity of Ethanol Extract and AgNPs

The antibacterial activity was studied by well diffusion method against microorganisms viz. *Escherichia coli*, *Bacillus cereus*, *Methicillin Resistance S. aureus* (MRSA) and *Staphylococcus aureus*.

2.8 Antifungal Activity of Ethanol Extract and AgNPs

The antifungal activity was studied by well diffusion method against the microorganism viz. *Aspergillus niger*, *Alternaria solani*, *Fusarium oxysporum* and *Penicillium chrysogenum*.

2.9 Antioxidant Activity of Ethanol Extract and AgNPs (Molyneux 2004; Williams et al. 1995)

2.9.1 Preparation of Various Concentration of Ethanol Extracts and Standard Drug

Ascorbic acid was taken as standard because it is highly antioxidant. Standard solution of ascorbic acid as well as ethanol extract were prepared viz. 50, 100, 200, 300, 400 and 500 μg/ml. AgNPs were evaluated for antioxidant activity.

2.10 Preparation of Test Sample and Standard Sample

DPPH was prepared in dark due to its high oxidizable nature. DPPH solution was prepared by weighing 20 mg and dissolved in 100 ml methanol. 3 ml of various concentration of test sample (ethanol extract/silver nanoparticles) and standard ascorbic acid solution were mixed separately with 1 ml of DPPH solution and incubated for 30 min. Sample absorbance was taken with the help of UV spectrophotometer at 517 nm.

Calculate the % activity of test sample and standard sample from the following formula:

$$\%\text{Antioxidant Activity} = \frac{\text{Absorbance of Control} - \text{Absorbance of Individual Concentration}}{\text{Absorbance of Control}} \times 100$$

3 Results and Discussion

The yield (grams) of ethanol extract of leaves was 5.91 gm. The ethanol extract of leaves of *Cucurbita pepo* was subjected to various qualitative phytochemical tests. The ethanol extract contains carbohydrate, alkaloids, phenolic compounds, protein and amino acids, saponins, sterols and terpinoids, fats and fixed oil which is among all tested phytoconstituents.

3.1 Synthesis of AgNPs

Ethanol extract was subjected for synthesis of AgNPs and there is reduction of Ag+ ions into silver particles during exposure to plant extract which is identified by change in the colour of the solution. AgNPs showed yellowish-brown colour in aqueous solution due to surface plasmon response in AgNPs. So when plant extract was added with aqueous solution of Ag+ ions, it started to change colour from light yellow to brown because of the reduction of Ag+ ions which indicates the formation of silver nanoparticles (Jain et al. 2009).

3.2 Observation of Colour Change

After synthesis of silver nanoparticles, the colour of the solution changed from light yellow to brown. Colour change of the solution is due to the presence of some chemical compounds such as alkaloids, carbohydrates, saponins, steroids, etc. which acts as reducing agent that reduce silver ion to silver atom (Fig. 1) (Niraimathi et al. 2013; Oluwaniyi et al. 2015; Panigrahi 2013).

The AgNPs in the solution were analysed by UV-visible spectrophotometer and appears peaks at near 460 nm (Pantidos and Horsfall 2014; Noginov et al. 2007).

3.3 UV-Visible Spectral Analysis

Nanoparticles show their characteristics optical properties because of surface plasmon resonance which depends on shape and size of particles and their refractive index. The change in colour of the solution due to reduction of silver ions by excitation of surface plasmon vibrations in silver nanoparticles which was monitored by UV-Visible spectrum and spectrum appears at near 460 nm which indicates the formation of silver nanoparticles (Fig. 2) (Swamy et al. 2015).

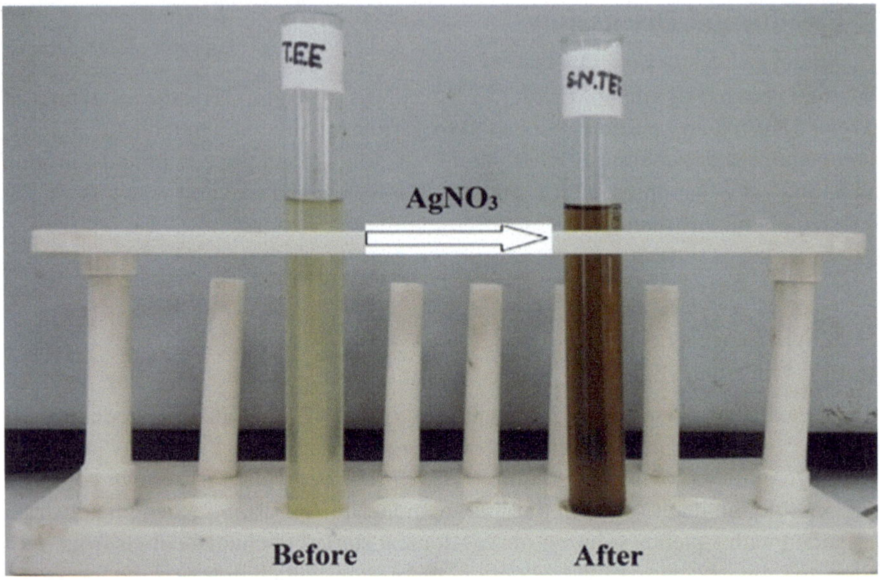

Fig. 1 Colour change of the solution after synthesis of silver nanoparticles

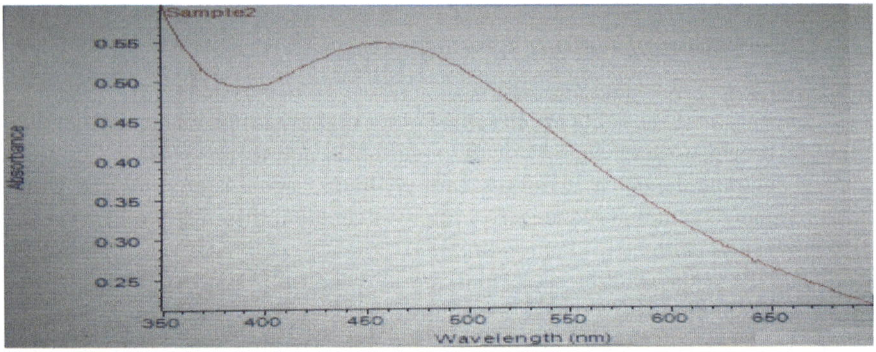

Fig. 2 UV-spectra of synthesize AgNPs

3.4 SEM Analysis

AgNPs were analysed by SEM shows in high-density AgNPs image. It was shown that AgNPs were formed with 5–100 nm diameter ranging. The SEM image of AgNPs indicates interactions between the various phytochemical molecules bound to the AgNPs. Images of biosynthesized nanoparticles in SEM show surface morphology of AgNPs were clearly indicated that AgNPs were roughly spherical in shape and uniformly distributed and agglomeration was found in AgNPs as shown in Figs. 3 and 4.

Fig. 3 SEM analysis of synthesize AgNPs

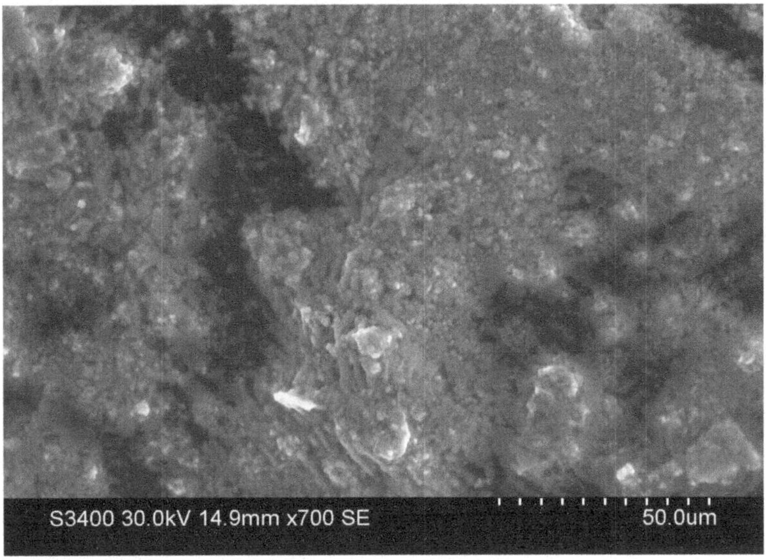

Fig. 4 SEM analysis of synthesize AgNPs

Acknowledgements The authors are thankful to the Management of DIBNS, Dehradun, for providing necessary facilities for completion of this work.

References

Badr SE, Shaaban M, Elkholy YM, Helal MH, Hamza AS, Masoud MS, Elsafty MM (2011) Chemical composition and Biological activity of ripe pumpkin fruits (Cucurbita pepo) cultivated in Egyptian habitats. Nat Prod Res 25(16):1524–1539

Barry AL, Fibiger (1976) The antimicrobial susceptibility test principle and practices. 180

Caili FU, Huan SHI. Quanhong LI (2006) A review on pharmacological activities and utilization technologies of pumpkin 73–80

D'britto V, Devi PP, Prasad BLV, Dhawan A, Mantri VG, Prabhune A (2012) Medicinal plant extracts used for blood sugar and obesity therapy shows excellent inhibition of invertase activity: synthesis of nanoparticles using this extract and its cytotoxic and genotoxic effects. Int J Life Sci Pharma Res 2(3):61–74

Goodsell DS (2004) Bionanotechnology: lessons from nature. Wiley, New York (2004)

Hettiarachchi MA, Wickramarachchi PASR (2011) Synthesis of chitosan stabilized silver nanoparticles using gamma ray radiation and characterization. J Sci Univ Kelaniya 6:65–75

Huang J, Li Q, Sun D, Lu Y, Su Y, Yang X et al (2007) Biosynthesis of silver and gold nanoparticles by novel sundried Cinnamomum camphora leaf. Nanotechnology 18:105104

Jain D, Kumar Daima S, Kachhwaha S, Kothari SL (2009) Synthesis of plant mediated silver nanoparticles using Papaya fruit extract and evaluation of their antimicrobial activities. Dig J Nanomater Biostructures 4(3):557–563

Martha R, Gutierrez P (2016) Medicinal chemistry review of Cucurbita pepo (Pumpkin) its phytochemistry and pharmacology 6:12–21

Mohamed GA, Ibrahim SRM, Sayed HM (2009) Phenolic constituents of Cucurbita pepo L Eskandrani (Summer Squash) Flower. Bull Pharm Sci 32:311–319

Molyneux P (2004) The use of the stable free radical diphenylpicrylhydrazyl (DPPH) for estimating antioxidant activity. Songklanakarin J 211–219

Nawirska-Olszanska A, Kita A, Biesiada A, Sokol-L-etowska A, Kucharska AZ (2013) Characteristics of antioxidant activity and composition of pumpkin seed oil in 12 cultivers. Food Chem 139(1):155–161

Niraimathi KL, Sudha V, Lavanya R, Brindha P (2013) Biointerfaces biosynthesis of silver nanoparticles using Alternanthera sessilis (Linn.) extract and their antimicrobial, antioxidant activities. Colloids Surf B: Biointerfaces 102:288–291

Noginov MA, Zhu G, Bahoura M, Adegoke J, Small C et al (2007) The effect of gain and absorption on surface plasmons in metal nanoparticles. Appl Phys B Lasers Opp 86:455–460

Oluwaniyi OO, Adegoke HI, Adesuji ET, Alabi AB, Bodede SO (2015) Biosynthesis of silver nanoparticles using aqueous leaf extract of Thevetia peruviana Juss and its antimicrobial activities. Appl Nanosci 6:903–912

Panigrahi T (2013) Synthesis and characterization of silver nanoparticles using leaf extract of Azadirachta indica. Dep Life Sci Natl Inst Technol Rourkela (Doctoral dissertation)

Pantidos N, Horsfall LE (2014) Biological Synthesis of Metallic Nanoparticles by Bacteria, Fungi and Plants. J Nanomed Nanotechnol 5:1

Parashar V, Parashar R, Sharma B, Avinash CP (2009) Parthenium leaf extract mediated synthesis of silver nanoparticle. Dig J Nanomater Biostructures 4(1):45–50

Rahman AHMM (2013) Ethno-medico botanical investigation on cucurbits of Rajshahi division. Bangladesh J Med. Plants Stud 1(3):118–125

Shankar SS, Ahmad A, Pasricha R, Sastry M (2003a) Bioreduction of chloroaurate ions by geranium leaves and its endophytic fungus yields gold nanoparticles of different shapes. J Mater Chem 13:1822–1826

Shankar SS, Ahmed A, Sastry M (2003b) Geranium leaf assisted biosynthesis of silver nanoparticles. Biotechnol Prog 19:1627–1631

Song JY, Kim BS (2008) Rapid biological synthesis of silver nanoparticles using plant leaf extracts. Bioprocess Biosyst Eng 32:79–84

Swamy MK, Sudipta KM, Jayanta K, Balasubramanya S (2015) The green synthesis, characterization, and evaluation of the biological activities of silver nanoparticles synthesized from Leptadenia reticulata leaf extract. Appl Nanosci 5:73–81

Tiwari P, Kumar B, Kaur M, Kaur G, Kaur H (2011) Phytochemical screening and extraction. Int Pharm Sci 1:98–106

Williams WB, Cuvelier ME, Berset C (1995) Use of a free radical method to evaluate antioxidant activity. Lebensm-Wiss U Technol 28:25–30

Methane Emission from Kaliveli Wetland: An Assessment

C. Luithui, Tasneem Abbasi and S. A. Abbasi

Abstract In the backdrop of great paucity of information on greenhouse gas emissions from Indian wetlands in general and wetlands of peninsular South India in particular, the present studies have been carried out. They pertain to Kaliveli lake—which has a water-spread of about 70 km^2 when full—situated between Chennai and Puducherry. The study reveals that, contrary to the expectation, the freshwater part of Kaliveli contributes less than 20 mg/m^2 h of methane flux in all the three seasons that occur in this region.

Keywords Kaliveli · Methane · Chennai · Puducherry

1 Introduction

1.1 Approach to the Present Study

Methane (CH_4) is the second most abundant greenhouse gas (GWG) after carbon dioxide (CO_2), accounting for 18% of the world's GWG radiative forcing (Abbasi and Abbasi 2011; IPCC 2013; Bharali et al. 2017). Of the prevailing CH_4 concentration in the atmosphere—183 parts per billion—which has been steadily increasing due to anthropogenic causes, 41% is estimated to be contributed by wetlands (Duchemin et al. 2006; Abbasi et al. 2011). Flooded paddy fields are among the major contributors but natural wetlands are also feared to contribute, especially due to their eco-degradation trigged by anthropogenic pollution (Abbasi and Chari 2008; Abbasi and Abbasi 2010a, 2018).

C. Luithui · T. Abbasi · S. A. Abbasi (✉)
Centre for Pollution Control & Environmental Engineering, Pondicherry University, Chinnakalapet 605014, Puducherry, India
e-mail: abbbasi.cpee@gmail.com

T. Abbasi
e-mail: tasneem.abbasi@gmail.com

N. A. Siddiqui et al. (eds.), *Advances in Water Pollution Monitoring and Control*, Springer Transactions in Civil and Environmental Engineering,
https://doi.org/10.1007/978-981-32-9956-6_14

In two recent reviews, focusing on India (Luithui et al. 2017) and peninsular South India (Dhanuja et al. 2018), we have shown that even as an increasingly vast corpus of information is in place, *vis a vis,* the GHG emissions from wetlands of several countries, information pertaining to India is shockingly sparse. This is even more so for peninsular South India even though some of India's largest paddy growing states, as also a large number of natural wetlands, occur in this region. Only a solitary study, on CH_4 emissions occurring in the Godavari estuary, has been conducted on the natural wetlands of this region (Dhanuja et al. 2018), by Krupadam et al. (2007). Even that study is 11-year old now. Its findings are at best historical and cannot be seen as representing the current picture.

In view of the near-total absence of information on GHG emissions from the wetlands of Southern India, we have carried out the present study on the Kaliveli lake, situated about 85 km South of downtown Chennai. It is one of the largest lakes of India, covering an area of about 70 km^2 when it is full by the end of the northwest monsoon.

1.2 The Kaliveli Wetland

Kaliveli (Fig. 1) is a large and shallow wetland, with an average depth of 90 cm when it is full. With a catchment spanning ~756 km^2, its petal-shaped water body lies parallel to the east coast (Fig. 1). Its southern tip is situated about 18 km north of downtown Puducherry and the northern tip is ~85 km south of downtown Chennai. A narrow channel links Kaliveli with the Bay of Bengal via Yedayanthitu estuary. A few salient features of Kaliveli are presented in Table 1.

Kaliveli fills up to its maximum extent by the end of the northeast monsoon (Fig. 2a), and then shrinks in the subsequent months (Fig. 2b). The freshwater part of the wetland dries out completely during the summer months of March–May (Fig. 2c). Thus, even as the wetland is normally full of freshwater in the aftermath of the north-east monsoon which comes during November–December—having received copious runoff from neighboring farmlands and other parts of the watershed—it begins to dry up rapidly as summer sets in by late February. Subsequently, as the run-in of freshwater diminishes, there is inflow of seawater from the estuary and the wetland becomes brackish at its northern end (Fig. 2d).

2 Materials and Method

The sampler used, the method of sampling, the estimation of the methane emission, and the associated analytical quality control were as detailed elsewhere (Abbasi 1998; Luithui 2018). All samples were drawn from a single location (Fig. 3).

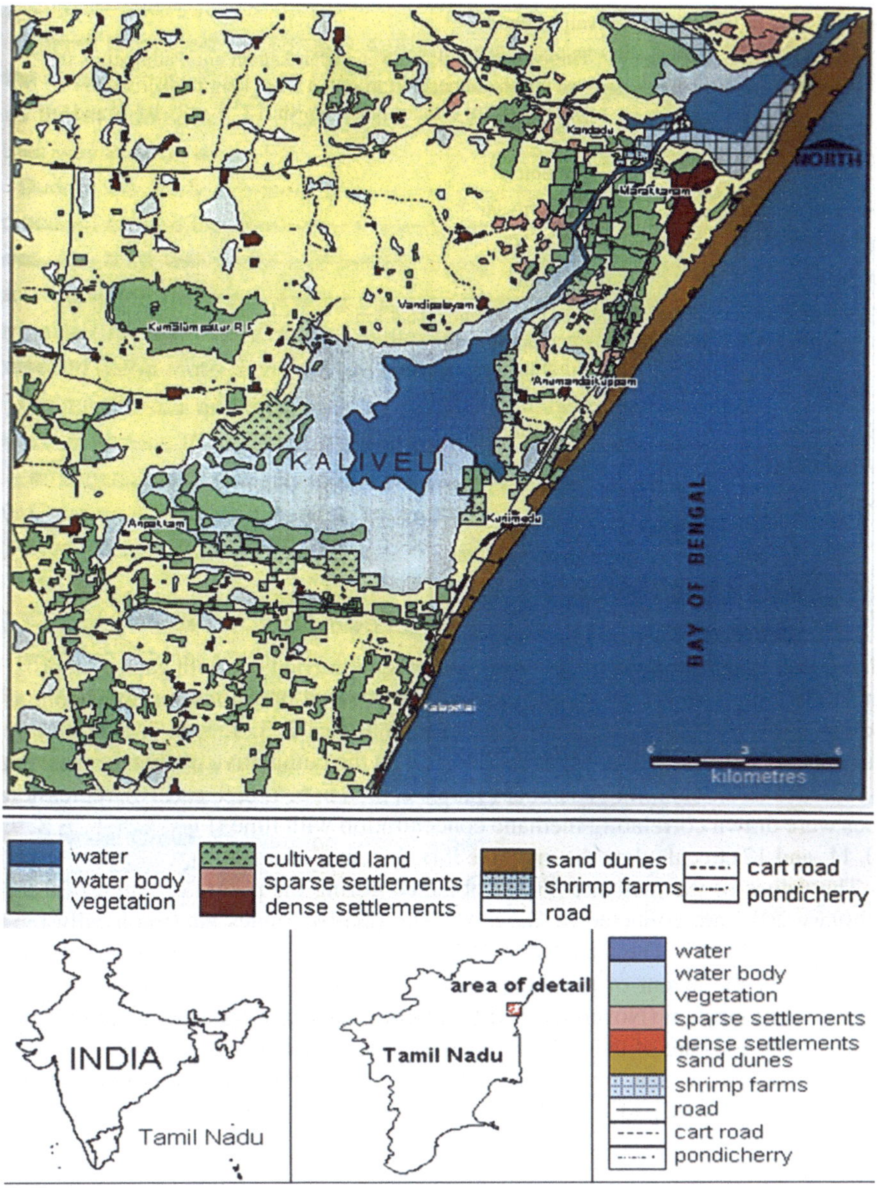

Fig. 1 Location and land use/land cover map of Kaliveli

Fig. 3 Location where the sampling was conducted

Table 2 Methane emission from Kaliveli lake on 19.08.2017

Time after setting the sampler (min)	Chamber air temperature (°C)	Water temperature (°C)	Methane concentration (mg/m^3)
20	35.4	36.3	2.8
40	38.7	34	3.4
60	38.5	34.5	4.01
80	39.4	35	3.8

Table 3 Methane emission from Kaliveli lake on 18.9.2017

Time after setting the sampler (min)	Chamber air temperature (°C)	Water temperature (°C)	Methane concentration (mg/m^3)
20	33.2	33.2	2.8
40	35.6	33.6	2.8
60	36.2	33.6	3.1
80	36.9	33.8	3.1
100	37.8	33.9	3.3

Table 4 Methane emission from Kaliveli lake on 7.10.2017

Time after setting the sampler (min)	Chamber air temperature (°C)	Water temperature (°C)	Methane concentration (mg/m^3)
20	33.4	32.7	2.2
40	35.6	32.9	2.7
60	36.0	32.9	2.9
80	35.8	33.0	2.9

Table 5 Methane emission from Kaliveli lake on 27.10.2017

Time after setting the sampler (min)	Chamber air temperature (°C)	Water temperature (°C)	Methane concentration (mg/m^3)
20	27.6	27.4	2.1
40	27.6	27.5	2.4
60	27.8	27.5	2.7
80	27.9	27.5	2.7

Table 6 Methane emission from Kaliveli lake on 16.11.2017

Time after setting the sampler (min)	Chamber air temperature (°C)	Water temperature (°C)	Methane concentration (mg/m^3)
20	26.4	27.2	2.2
40	26.4	27.2	2.5
60	26.5	27.0	2.5
80	26.4	27.3	2.7

Table 7 Methane emission from Kaliveli lake on 6.12.2017

Time after setting the sampler (min)	Chamber air temperature (°C)	Water temperature (°C)	Methane concentration (mg/m^3)
20	27.3	27.1	2.1
40	27.5	27.3	2.2
60	27.8	27.2	2.1
80	27.4	27.4	2.4

Table 8 Methane emission from Kaliveli lake on 26.12.2017

Time after setting the sampler (min)	Chamber air temperature (°C)	Water temperature (°C)	Methane concentration (mg/m^3)
20	28.4	27.8	2.2
40	29.0	27.8	2.4
60	30.0	27.9	2.4
80	31.4	27.9	2.6

Table 9 Methane emission from Kaliveli lake on 15.01.2018

Time after setting the sampler (min)	Chamber air temperature (°C)	Water temperature (°C)	Methane concentration (mg/m^3)
20	29.4	27.5	2.4
40	31.2	27.7	2.2
60	32.3	27.8	2.4
80	32.7	27.7	2.5

Table 10 Methane emission from Kaliveli lake on 5.2.2018

Time after setting the sampler (min)	Chamber air temperature (°C)	Water temperature (°C)	Methane concentration (mg/m^3)
20	28.6	28.8	2.4
40	29.4	29.0	2.5
60	31.1	29.1	2.5
80	32.8	29.0	2.8

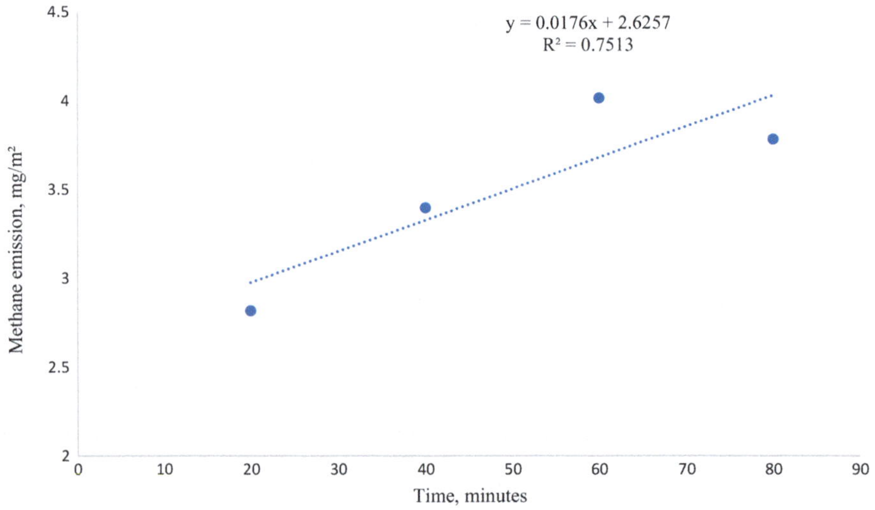

Fig. 4 Regression correlation between methane concentration and time; cf Table 2

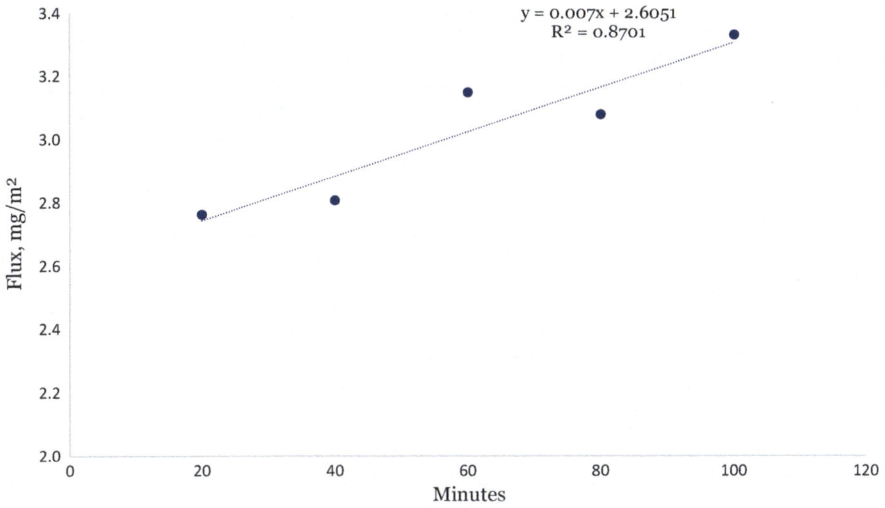

Fig. 5 Regression correlation between methane concentration and time; cf Table 3

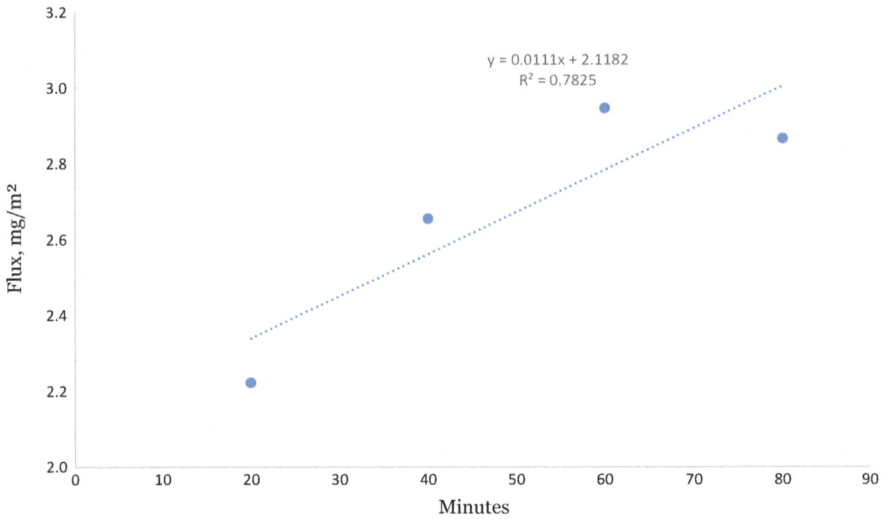

Fig. 6 Regression correlation between methane concentration and time; cf Table 4

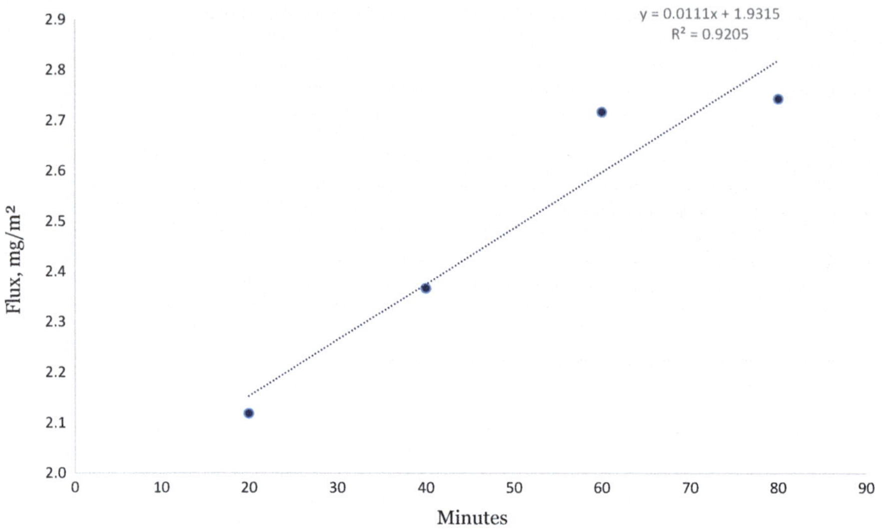

Fig. 7 Regression correlation between methane concentration and time; cf Table 5

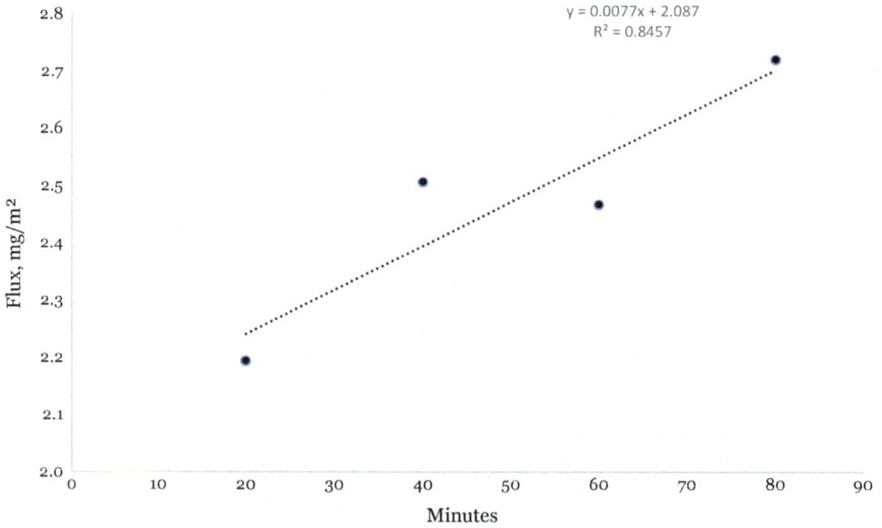

Fig. 8 Regression correlation between methane concentration and time; cf Table 6

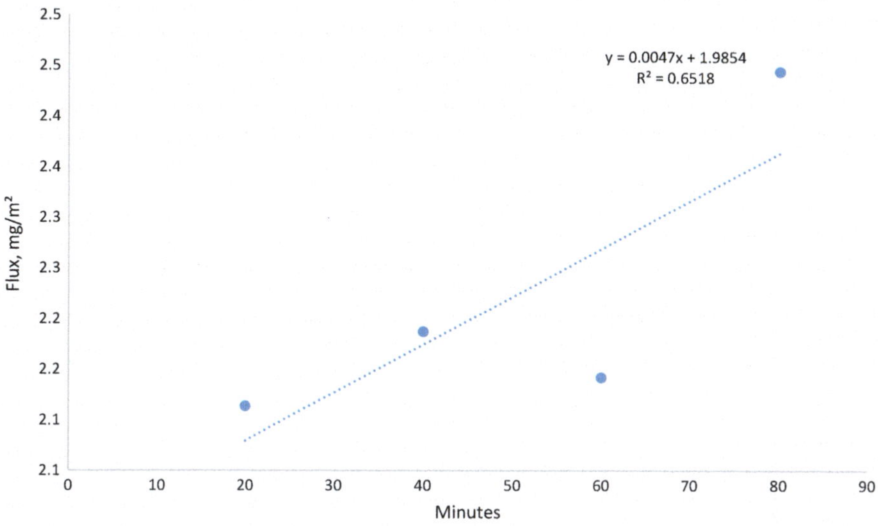

Fig. 9 Regression correlation between methane concentration and time; cf Table 7

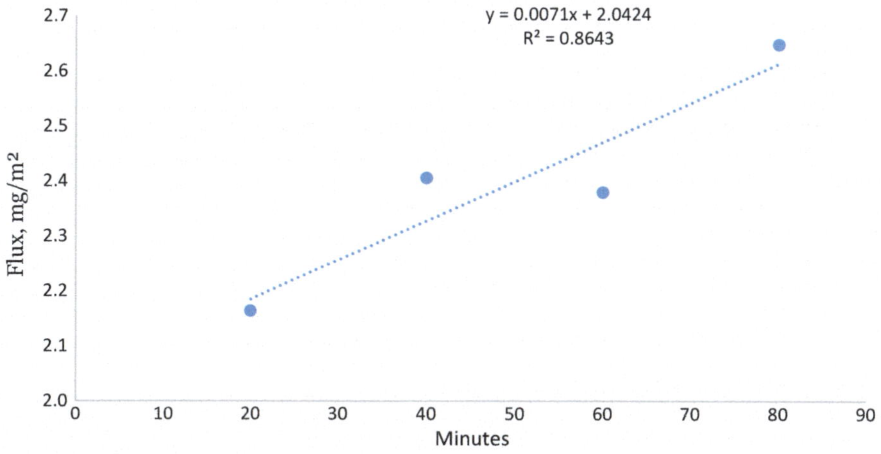

Fig. 10 Regression correlation between methane concentration and time; cf Table 8

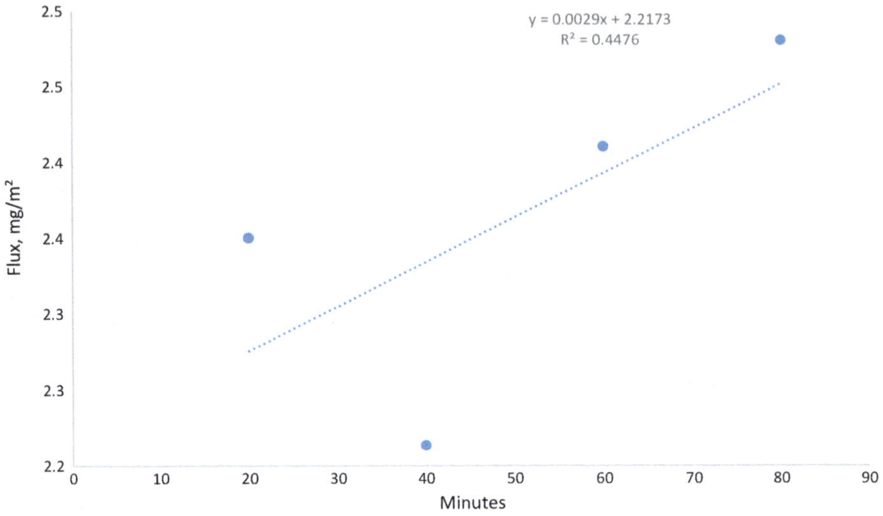

Fig. 11 Regression correlation between methane concentration and time; cf Table 9

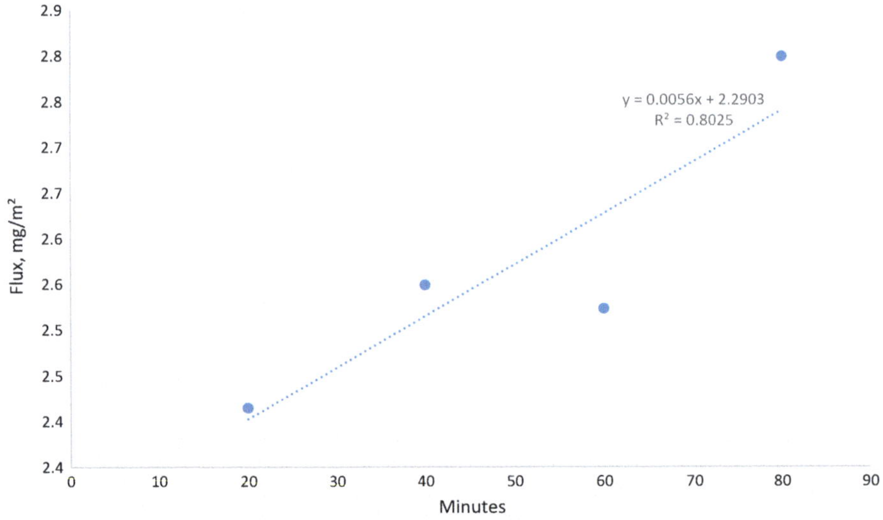

Fig. 12 Regression correlation between methane concentration and time; cf Table 10

Table 11 Methane emission from Kaliveli lake during different months

Sampling dates	Average temperature (°C)	Methane emissions (μg/m^2 h)
19.08.2017	38.6	20.0
18.9.2017	35.9	8.0
7.10.2017	35.2	13.0
27.10.2017	27.7	13.0
16.11.2017	26.4	9.0
6.12.2017	27.5	6.0
26.12.2017	29.7	9.0
15.01.2018	31.4	3.0
5.2.2018	30.4	7.0

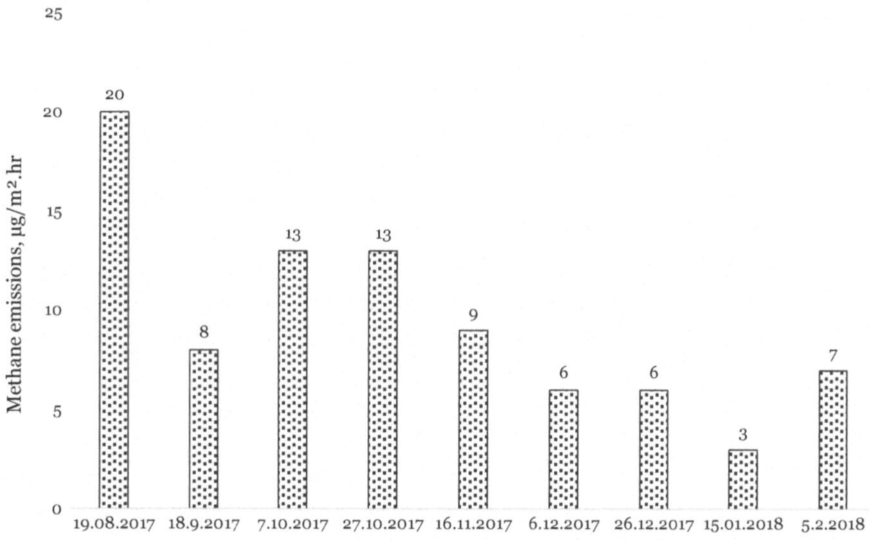

Fig. 13 Methane emission from Kaliveli lake during different months; cf Table 11

may, however change if the wetland becomes eutrophic due to anthropogenic tempering, leading to infestation of aquatic weeds (Abbasi and Nipany 1986; Abbasi and Abbasi 2010b).

Acknowledgements SAA thanks the Council of Scientific and Industrial Research (CSIR), New Delhi, for the Emeritus Scientist grant (21(1034)/16/EMR-II) which enabled him to oversee and document this study.

References

Abbasi SA (1998) Water Quality – sampling and analysis discovery publishing house New Delhi, viii + 212 pp

Abbasi T, Abbasi SA (2010a) Remote sensing, GIS and wetland management. Discovery Publishing House, New Delhi vii + 411 pp

Abbasi T, Abbasi SA (2010b) Production of clean energy by anaerobic digestion of phytomass—New prospects for a global warming amelioration technology. Renew Sustain Energ Rev 14(6):1653–1659

Abbasi T, Abbasi SA (2011) Decarbonization of fossil fuels as a strategy to control global warming. Renew Sustain Energ Rev 15(4):1828–1834

Abbasi T, Abbasi SA (2018) Perspectives in pollution control and sustainable development. Discovery Publishing House, New Delhi, xi + 569 pp

Abbasi SA, Chari KB (2008) Environmental management of urban lakes. Discovery Publishing House, New Delhi, Viii + 269 pp

Abbasi SA, Nipaney PC (1986) Infestation by aquatic weeds of the fern genus: its status and control. Environ Conserv 13(3):235–241

Abbasi T, Tauseef SM, Abbasi SA (2011) Global warming and the role of wetlands. Lap Lambart Academic, Germany; ISBN 978-3846556009. Xiv + 264 pp

Bharali A, Baruah KK, Gogoi N (2017) Methane emission from irrigated rice ecosystem: relationship with carbon fixation, partitioning and soil carbon storage. Paddy Water Environ 15(2)

Dhanuja C, Abbasi T, Abbasi SA (2018) Greenhouse gas emissions from paddy fields in peninsular South India. Int J Eng Sci Res 6:56–63

Duchemin E, Lucotte M, Canuel R, Nicolas S (2006) First assessment of methane and carbon dioxide emissions from shallow and deep zones of boreal reservoirs upon ice break-up. Lakes Reserv Res Manag

IPCC (2013) Climate change 2013: the physical science basis. Contribution of working group I to the fifth assessment report of the intergovernmental panel on climate change [Stocker TF, Qin D, Plattner G-K, Tignor M, Allen SK, Boschung SK, Nauels A, Xia Y, Bex V, Midgley PM (eds)]. Cambridge University Press, Cambridge, United Kingdom and New York, NY, USA, 1535 pp

Krupadam RJ, Ahuja R, Wate SR, Anjaneyulu Y (2007) Forest bound estuaries are higher methane emitters than paddy fields: a case of Godavari estuary, east coast of India. Atmos Environ 41(23):4819–4827

Luithui C (2018) Studies on greenhouse gas emissions from wetlands. PhD Thesis, Pondicherry University, p 217

Luithui C, Dhanuja C, Abbasi T, Abbasi SA (2017) Greenhouse gas emissions and wetlands: a brief review of the state-of-the-art. Int J Environ Sci Technol 8(2):01–50

Biosorption of Safranin-O Dye by Shell of Arachis Hypogaea

Varsha Panchal and Arpita Ghosh

Abstract Textile industries annually use nearly 10,000 of different dyes for various purpose of coloration. Nearly 25% of dyes are consumed in coloration and the rest amount discharged in effluent. This practice is making the surface water unfit for drinking, household purpose, and very dangerous for aquatic lives. In the present study, shell of *Arachis hypogaea* (peanut hull) was used to remove safranin-o dye (C.I. 50240) from synthetic aqueous solution. Effect of various parameters (pH, dye concentration, absorbent dose, time, and adsorbent size) was studied on the biosorption process of dye. The maximum removal (97%) was observed using 100 mg/l dye concentration using 7 g dose of biosorbent at pH 7 and 40 °C temperature. The study indicated that peanut hull (common solid waste from the household) can be used for treatment of textile effluent, which will also achieve the sustainability of waste management.

Keywords Biosorption · Safranin-O dye · Peanut hull · Textile · Waste management

1 Introduction

Water is essential for the survival of all living organism. It is a big dramatic concern that is increasing day by day because of many aspects of textile, leather, cosmetic, and food industries. These industries discharge different contaminants (such as dyes, metals, soap, detergents, and chemicals) widely used in manufacturing of different products (Panswed and Wongchaisuwan 1986; Ghosh et al. 2017a; Ghosh and Das 2014). The polluted water has been reported to be very toxic for aquatic life and also affects the food chain of natural water bodies (rivers, lakes). Conventionally in ETP, effluents are used to treat by primary treatment (coagulation, flocculation, and sedimentation), secondary treatment (bioreactors), and tertiary treatment (filtration,

V. Panchal · A. Ghosh (✉)
Department of Biotechnology, Manav Rachna International Institute Research and Studies, Faridabad, India
e-mail: arpi.335@gmail.com

N. A. Siddiqui et al. (eds.), *Advances in Water Pollution Monitoring and Control*, Springer Transactions in Civil and Environmental Engineering,
https://doi.org/10.1007/978-981-32-9956-6_15

air scouring). It is difficult to remove/degrade synthetic dyes due to their complex structure and nature (Ghosh et al. 2017b).

Dyes cause hazardous effect by blocking the sunlight that reduces photosynthesis or decreases the level of dissolved oxygen that becomes life-threatening for aquatic plants and animals (Bharati and Rames 2013). Dyes are also reported to be toxic, carcinogenic, and mutagenic (Namasivayam and Kavitha 2002; Chuah et al. 2005; Rizzuti and Lancaster 2013). Safranin-o (cationic dye, C.I. no. 50240) widely used in textile and pharmaceutical industries are known to be a cause of several health problems like irritating the respiratory system, skin infection, allergies, eye irritation, and digestive tract infection when ingested (Mohammed et al. 2014). Therefore, it is essential to decolorize the wastewater for making it safe and useable.

Different agro-waste materials such as coir pith, rice husk, soyabean hull, mango seed, corn husk, and banana peel were used to actively remove dye from wastewater (Namasivayam and Kavitha 2002; Chuah et al. 2005; Rizzuti and Lancaster 2013; Senthil Kumar et al. 2014; Paşka et al. 2014; Amela et al. 2012). Agro-waste materials are the good agents for treating the organic (proteins, fats, carbohydrates, dyes, soaps, detergents, and pesticides) and inorganic (acids, alkali, cations, and anions) compounds. These pollutants may change the buffer system and affect the pH of water bodies. Adsorption phenomenon is to attract and retain the pollutant on the surface of a substance to remove the pollutant. On the economic point of view biosorbent should be cheap, widely available, and eco-friendly and also could be reused. In this way, sustainable waste management can also be achieved. Table 1 is showing usage of different adsorbents for removal of safranin dye by different researchers.

In the recent study, peanut hull was used as a non-conventional adsorbent for removing safranin-o dye from aqueous solution. Peanut hull has high efficiency to remove the dyes such as methylene blue, neutral red, and brilliant cresyl blue reported by Gong et al. (2005) because of the presence of various structural components (28.8% lignin, 37% cellulose, 25% carbohydrates, and 8.2% protein) (Gong et al. 2005). Generally, huge amount of peanut hull as waste is generated from household in

Table 1 Safranin dye removal using different adsorbents

Adsorbents	Adsorbate	Adsorption capacity	Reference
Pineapple peel	Safranin-O	Specific uptake of dye: 21.7 mg/g	Mohammed et al. (2014)
Micellar enhance ultra violet	Safranin-T	Decolorization: 99%	Zaghbani et al. (2008)
NaOH treated rice husk	Safranin	Specific uptake of dye: 37.97 mg/g	Chowdhury et al. (2012)
Pretreated rice husk	Safranin and Methylene Blue	Specific uptake of dye: 45.582 mg/g	Chowdhury and Das (2011)
Alkali treated mango seed	Safranin	Specific uptake of dye: 31 mg/g	Malekbala et al. (2012)

Winter at north-India, which contributes to the high amount of woody waste-material that is also necessary to clean.

2 Materials and Methods

2.1 Dye Solution

Table 2 is showing IUPAC name, molecular weight, and chemical formula of safranin-o dye (Sahu and Patel 2016).

The dye was obtained from Thermo Electron India Pvt. Ltd., Mumbai-400710. Stock solution 1000 ml was prepared by dissolving 1.0 g dye in 1 L distilled water. The normal pH of the stock solution was around 6.

2.2 Biosorbent Sample

Peanut hull (shell of *Arachis hypogaea* L) was collected from the nearby local market at Dabua Chowk, Faridabad. Then peanut hull rinsed with distilled water 2–3 times to remove dust and water-soluble impurities from its surface and dried in hot air oven at 50 °C for 48 h until the hull becomes crispy. After that, the sample was crushed to 1 mm particle size and stored in air-tight container. This powdered material was further used as an adsorbent in the biosorption process of safranin-o dye.

2.3 Instruments and Chemicals

The solutions of HCl (Thermo Electrons LLS India Pvt. Ltd., Mumbai) and NaOH (Qualigen fine chemicals, Mumbai) were used to set different ranges of pH of dye solution through pH meter. Incubator Shaker (Biosphere Corporation) was carried out for the experiment (biosorption). Separation of adsorbent particles from dye solution was done by centrifugation (RM-12C Bench Top Centrifuge by REMI Motors Limited). The amount of removal of dye from aqueous solution after biosorption was analyzed via Delux Photo Colorimeter manufactured by Labrotonics panchkula, India.

Table 2 Characteristics of safranin-o dye

IUPAC name	3,7 diamino 2,8 dimethyl 5 phenyl phenazarium chloride
M.W	35.88 per mol
Chemical formula	$C_{20}H_{19}CLN_4$

2.4 *Method*

The experiment planned to determine the efficiency of the peanut hull for removal of safranin-o dye from aqueous solution. Stock solution (1000 mg/l) of safranin-o was prepared by dissolving 1 g of dye in 1 L. The dye solutions (100 ml solution in 250 ml conical flask) of different concentrations were prepared by diluting the stock solution in distilled water. The fixed quantity of adsorbent was added to each flask and kept in incubator shaker at 125 rpm for 4–5 h at 30 °C. In each hour, sample solution was collected in centrifuge tube and labeled serial wise one by one. Then centrifugation was performed on the sample for 15 min at 2000 rpm. After that optical density was determined through colorimeter. The % removal of dye from solution was calculated by the following formula (1):

$$\%R = 100 * \left\{ \frac{(C_0 - C_e)}{C_0} \right\} \tag{1}$$

where R is % removal of dye from aqueous solution, C_0 is initial dye concentration, and C_e is the final concentration.

The used adsorbent samples (from all experiments) were dried for 24 h for desorption and reuse study. This dried adsorbent sample (1 g) was soaked in 0.2 (M) HCl solution and another sample was soaked in 0.2 (M) solution of NaOH for removing the adsorbed dye, so that adsorbent can be reused. Then the sample again was rinsed with distilled water 2–3 times and dried again. To check the removal tendency by that sample dye solution was prepared of 100 mg/l. After that, both desorbed samples of biomass were added in each conical flask with 7 pH and kept under incubator shaker for 4 h. In each hour, samples were collected and centrifuged at 2000 rpm for 15 min. Then optical density was observed through colorimeter.

3 Results and Discussion

3.1 *Effect of pH*

The pH of solution plays an important role in adsorption because it affects the buffer system (Sonawane and Shrivastava 2011). To study the effect of pH on adsorption experiment were using 100 mg/l dye solution in 250 ml conical flask. The pH of the dye solution was adjusted at 3, 5, 7, and 9. Acidic pH was adjusted using 0.1 M HCl and basic pH were via using 0.1 M NaOH dropwise. Then 1 g adsorbent was added in each flask and kept on incubator shaker for 4 h. The samples were collected after each hour and centrifuged. Then the supernatant of samples were analyzed through colorimeter and % removal was determined. Plotted graph at x-axis time ranging in minutes (60, 120, 180, and 240) and Y-axis represents % removal. Maximum removal

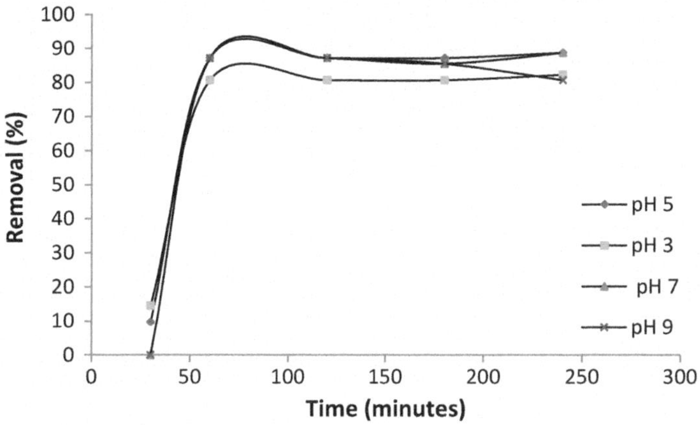

Fig. 1 The effect of pH on uptake of safranin-o using adsorbent dose 1 g at agitation speed 125 rpm

of safranin-o dye was observed at pH 7 but after 3 h desorption also appeared in pH 9 (Fig. 1).

3.2 Effect of Adsorbent Dose

Adsorbent dose majorly affects the adsorption process. As the adsorbent dose increases the removal of dye also increases. To study the effect of adsorbent dose various concentration of adsorbent dose was added ranging from 3, 5, and 7 g in 100 mg/l dye solution at pH 7, 30 °C temperature and kept in incubator shaker for 4 h. The optical density checked through colorimeter and % removal was calculated by above given Eq. (1). Maximum removal was observed using 7 g of adsorbent dose because more area of biosorbent was available for adsorption (Fig. 2).

3.3 Effect of Dye Concentration

For the study of dye concentration, different dye solution was prepared with different mg/l 10, 25, and 50 in each dye solution fixed quantity (1 g) of adsorbent were added and kept in incubator shaker for 4 h at 30 °C temperature. The pH of dye solution was set 7. The removal efficiency of dye from solution was observed by colorimeter. Maximum removal was observed at 25 mg/l solution but after sometime desorption was also analyzed (Fig. 3).

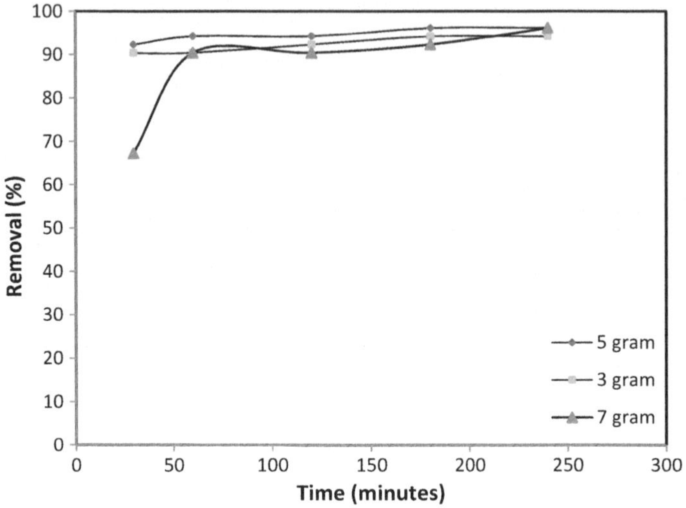

Fig. 2 The effect of adsorbent dose on removal of safranin-o using adsorbent dose 7 g, agitation speed 125 rpm at pH 7

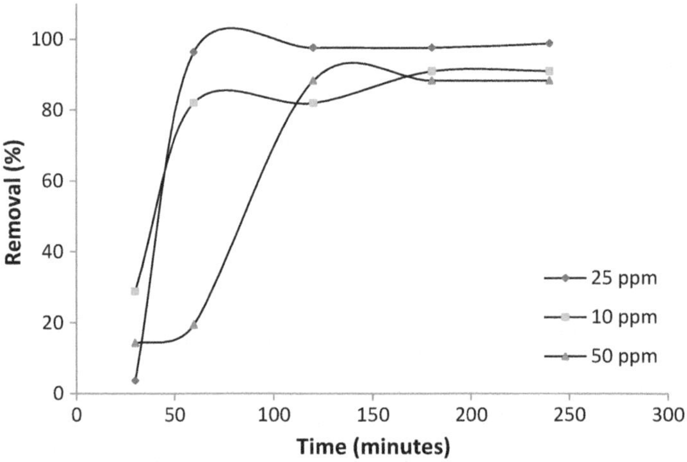

Fig. 3 Maximum removal of safranin-o dye using adsorbent dose 1 g at 25 mg/l solution and agitation speed 125 rpm at pH 7

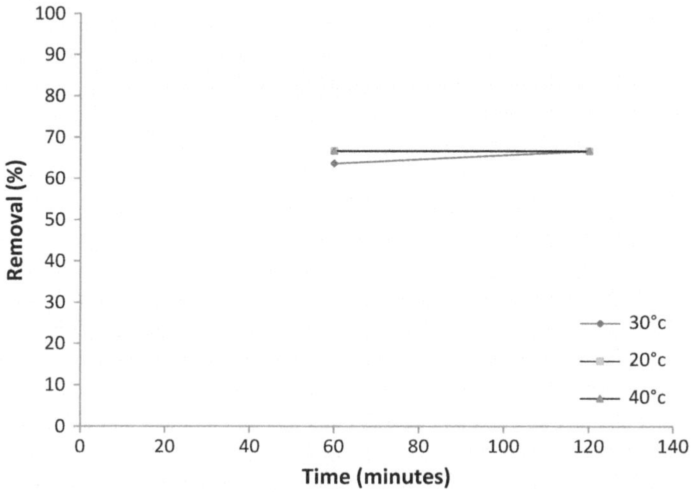

Fig. 4 The effect of temperature on uptake of safranin-o using adsorbent dose 1 g, pH 7

3.4 Effect of Temperature

On the basis of temperature, it's very difficult to analyze the removal of dye by adsorbent. Temperature is the only parameter that specifies that the process is exothermic or endothermic. Temperature also gives the information change in enthalpy and entropy during adsorption. The temperature was varied as 20, 30, and 40 °C for 2 h using the dye solution of 100 mg/l at pH 7. The removal efficiency of dye was the same at 20 and 40 °C but slightly showed difference at 30 °C (Fig. 4).

3.5 Desorption and Reuse of Biosorbent

Desorption is the opposite process of adsorption. Desorption helps to understand the mechanism of adsorption that's why it is very important to study. Desorption actively seen in basic dyes as compared to reactive dyes. In this experiment, all the parameters are done in pH 7, i.e., a neutral pH. In neutral pH bond between adsorbent (peanut hull) and sorbent (dye solution) gets weak or breaks easily. After desorption that adsorbent is reused for adsorption. Maximum reuse efficiency was observed as 71% in case of NaOH and 25% in case of HCl after desorption (Fig. 5).

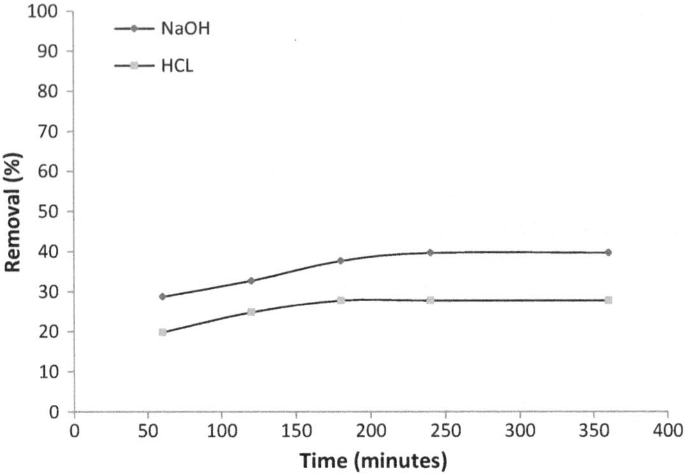

Fig. 5 The uptake efficiency of removal using adsorbent dose 1 g, pH 7, agitation speed 125 rpm

4 Conclusion

The present study shows very good results of adsorption of safranin-o dye by peanut hull. Peanut hull effectively removes dye from aqueous solution. The maximum removal obtained is 97%. Complete removal of dye could be done by providing appropriate amount of adsorbent dose, pH, and temperature for cleaning the wastewater. It could be a very successful treatment for treating the wastewater. Peanut hull could be widely available for cleaning the dye from polluted water. The study indicated that peanut hull (solid waste from household) can be used for the treatment of textile effluent, which will help to achieve the sustainability of waste management.

References

Amela K, Hassen MA, Kerroum D (2012) Isotherm and kinetics study of biosorption of cationic dye onto banana peel. Energy Procedia 19:286–295

Bharati KS, Rames ST (2013) Removal of dye using agricultural waste as low cost adsorbents review

Chowdhury S, Das P (2011) Linear and nonlinear regression analyses for binary sorption kinetics of methylene blue and safranin onto pretreated rice husk. Bioremediat J 15(2):99–108

Chowdhury S, Mishra R, Kushwaha P, Saha P (2012) Removal of safranin from aqueous solutions by NaOH-treated rice husk: thermodynamics, kinetics and isosteric heat of adsorption. Asia-Pac J Chem Eng 7(2):236–249

Chuah TG, Jumasiah A, Azni I, Katayon S, Choong ST (2005) Rice husk as a potentially low-cost biosorbent for heavy metal and dye removal: an overview. Desalination 175(3):305–316

Ghosh A, Das P (2014) Optimization of copper adsorption by soil of polluted wasteland using response surface methodology. Indian Chem Eng 56(1):29–42

Ghosh A, Dastidar MG, Sreekrishnan TR (2017a) Bioremediation of chromium complex dyes and treatment of sludge generated during the process. Int Biodeterior Biodegrad 119:448–460

Ghosh A, Dastidar MG, Sreekrishnan TR (2017b) Response surface optimization of bioremediation of acid black 52 (Cr complex dye) using *Aspergillus tamarii*. Environ Technol 38(3):326–336

Gong R, Li M, Yang C, Sun Y, Chen J (2005) Removal of cationic dyes from aqueous solution by adsorption on peanut hull. J Hazard Mater 121(1–3):247–250

Malekbala MR, Soltani SM, Yazdi SK, Hosseini S (2012) Equilibrium and kinetic studies of safranine adsorption on alkali-treated mango seed integuments. Int J Chem Eng Appl 3(3):160

Mohammed MA, Ibrahim A, Shitu A (2014) Batch removal of hazardous safranin-O in wastewater using pineapple peels as an agricultural waste based adsorbent. Int J Environ Monit Anal 2(3):128–133

Namasivayam C, Kavitha D (2002) Removal of Congo Red from water by adsorption onto activated carbon prepared from coir pith, an agricultural solid waste. Dyes Pigm 54(1):47–58

Panswed T, Wongchaisuwan S (1986) Mechanisms of dye wastewater colour removal by magnesium carbonate-hydrated basic. Water Sci Technol 18(3):139–144

Paşka OM, Păcurariu C, Muntean SG (2014) Kinetic and thermodynamic studies on methylene blue biosorption using corn-husk. RSC Adv 4(107):62621–62630

Rizzuti AM, Lancaster DJ (2013) Utilizing soybean hulls and rice hulls to remove textile dyes from contaminated water. Waste Biomass Valoriz 4(3):647–653

Sahu MK, Patel RK (2016) Correction: Removal of safranin-O dye from aqueous solution using modified red mud: kinetics and equilibrium studies. RSC Adv 6(30):25472–25472

Senthil Kumar P, Palaniyappan M, Priyadharshini M, Vignesh AM, Thanjiappan A, Sebastina Anne Fernando P, Srinath R et al (2014) Adsorption of basic dye onto raw and surface-modified agricultural waste. Environ Prog Sustain Energy 33(1):87–98

Sonawane GH, Shrivastava VS (2011) Removal of hazardous dye from synthetic textile dyeing and printing effluents by Archis hypogaea L. shell: a low cost agro waste material. Desalin Water Treat 29(1–3):29–38

Zaghbani N, Hafiane A, Dhahbi M (2008) Removal of Safranin T from wastewater using micellar enhanced ultrafiltration. Desalination 222(1–3):348–356

Assessment of Heavy Metal and Physiochemical Parameter in Surface and Groundwater Quality of Dehradun District of Uttarakhand

Kanchan Deoli Bahukhandi, Nihal Anwar Siddiqui, S. K. Bartarya and Arora Shefali

Abstract Water pollution is one of the biggest environmental problems in India. Industrialization and urbanization had increased contamination in the water bodies. Over past few decades, the improper disposal of hazardous waste and electronic waste and toxic effluent had increased heavy metal contamination and physicochemical parameter in surface and groundwater bodies. The objectives of this research paper are to assess the heavy metals and physicochemical parameter in surface and groundwater in different land use pattern, i.e. industrial area, agricultural area and urban area. The surface water samples were collected from different streams and rivers and groundwater samples were collected from tubewell, well and borewells during summer season, pre-monsoon season and post-monsoon season. The parameters like pH, DO, temperature, conductivity, bicarbonate, fluoride and heavy metals Fe, Pb, Cu, Zn, Cd, Cr, As, Sr and B were analyzed in the laboratory.

Keywords Water quality · Industrial area · Urban area

1 Introduction

Water is the elixir of life. It is an essential component of all animal and plant life. Water is very precious natural resource present on earth. The ocean contains about 97% of earth's water resource which is unfit for human consumption and other uses because of high-salt content. Of the 3% freshwater, about 2% lies in the polar region and only about 1% is available for human consumption (Fig. 1).

The rapid industrialization and urbanization had led to the over-exploitation of natural resources. Industrial, domestic and agricultural activities use large quantities of water and all waste water discharged into water bodies. The anthropogenic activities are the main cause for a different kind of surface and groundwater pollution. The excess use of water, deforestation, urbanization and industrialization caused depletion and deterioration of surface and groundwater resources. Safe drinking water is

K. D. Bahukhandi (✉) · N. A. Siddiqui · S. K. Bartarya · A. Shefali
University of Petroleum and Energy Studies, Dehradun, India
e-mail: kanchan@ddn.upes.ac.in

N. A. Siddiqui et al. (eds.), *Advances in Water Pollution Monitoring and Control*,
Springer Transactions in Civil and Environmental Engineering,
https://doi.org/10.1007/978-981-32-9956-6_16

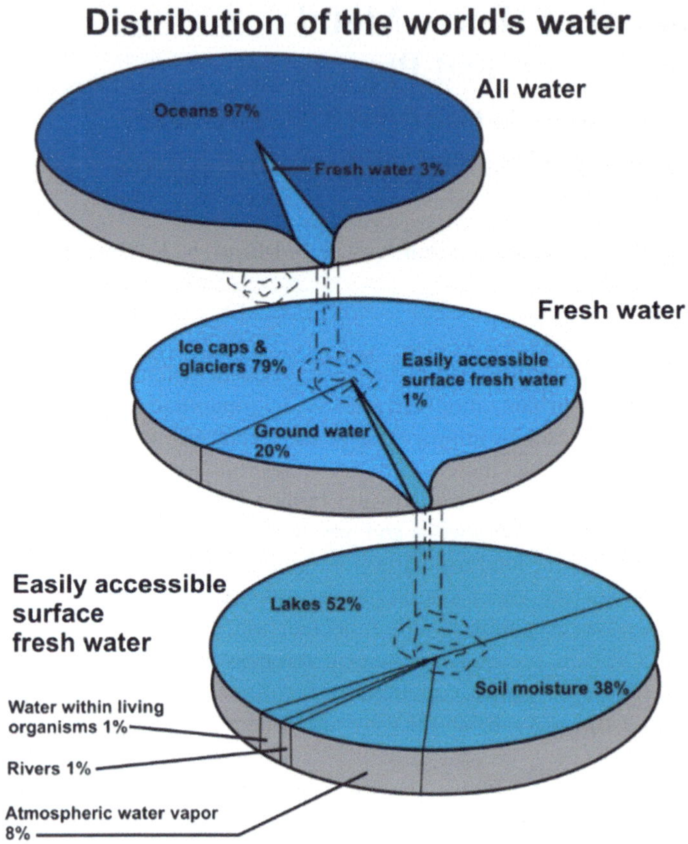

Fig. 1 Distribution of the world's water

the demand of the world to lead a healthy life. It has become imperative to protect and conserve the water of the world not only as a resource for their present use but also for their prospective uses.

The study on major ion chemistry and factor controlling ionic variation had been carried out by number of researchers in past by Bahukhandi et al. (2017, 2018), Bahukhandi and Aaron (2017), Kanchan et al. (2015), Bahukhandi and Bartarya (2012, 2014), Bartarya and Bahukhandi (2012), Bahukhandi and Siddiqui (2008), Raymahasay (1970), Abbas and Subramanain (1984), Dalai et al. (2002a, b), Krishnaswami et al. (1992), Pande et al. (1994), Subudhi et al. (2007), Chakrapani et al. (1995), Chakrapani (2002, 2005), Chakrapani and Subramanian (1996), Bhatt and Saklani (1996), Pandey et al. (1991), Galy and France-Lanord (1999), Meybeck (1976), Livingstone (1963) and APHA (1992), Gibbs (1970). Analysis of heavy metal and the associated physiochemical parameter is essential to understand the impact of water pollution on human health and developing mitigation measures in order to conserve and preserve water quality.

2 Methodology

The methodology consists of field survey, identification of sampling location, collection of surface water samples and groundwater samples, analysis of the samples in the field as well as in the laboratory. The surface water samples were collected from different rivers, i.e. Bindal River, Rispana River, Song River, Asan River and Suswa River while groundwater samples were collected from borewells, tube well and handpump in different lands use pattern, i.e. industrial area, agricultural area and urban area. Total 50 water samples were collected from surface and groundwater of Dehradun district. The parameters like pH, temperature EC and DO are measured at field while the measurement of heavy metals like Fe, Pb, Cu, Zn, Cd, Cr, As, Sr, B, all the samples were brought out in the laboratory using instruments like Inductive Coupled Plasma (ICP) and Atomic Absorption Spectrophotometer (AAS). Bicarbonate was measured by titration with 0.1 NHCL using methyl orange indicator. Nitrate and phosphate were measured with the help of ion chromatograph. The technique and method for collection and analysis of water sample were carried out as per American Public Health Association (APHA 1992).

3 Result and Discussion

3.1 *Spatial Variation in Water Quality*

3.1.1 Temperature

The average temperature in industrial area of groundwater and surface water of Dehradun district was 24 °C in Winter season, 26 °C in Summer and post-monsoon season. In urban area of groundwater of Dehradun, the mean temperature was 20 °C, 24 °C and 26 °C, respectively, in Winter, Summer and post-monsoon season (Fig. 2).

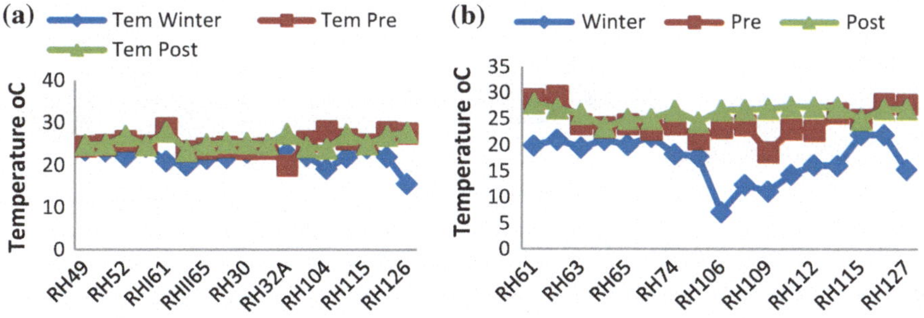

Fig. 2 **a** Temperature groundwater. **b** Temperature in surface water

3.1.2 pH

In urban area pH tends to be alkaline in nature with an average pH remained at 7.6 and 8 during Summer and post-monsoon season and does not show much changes in Winter season possibly due to the higher dissolution of carbonate rocks during post-monsoon season (Fig. 3).

3.1.3 Electric Conductivity

The mean value of EC in surface water and groundwater were ranged from 469 μs/cm to 816 μs/cm in all three seasons, i.e. Summer season, pre-monsoon season and post-monsoon in season (Fig. 4).

3.1.4 Bicarbonate

In groundwater of industrial area of Dehradun district (Selaqui) the average HCO_3 concentration varied from 78 mg/l to 277 mg/l in all seasons (Fig. 5).

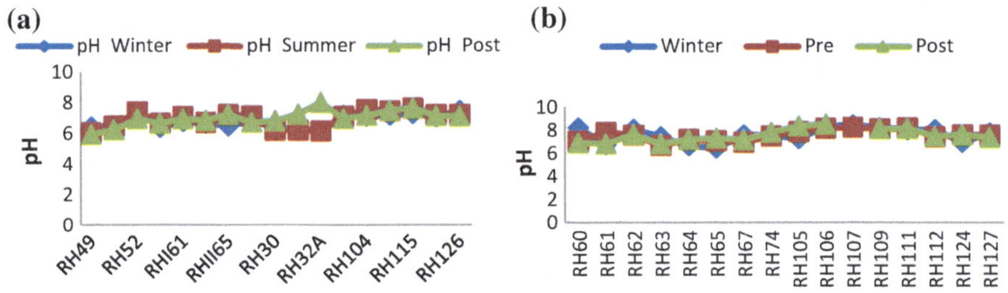

Fig. 3 **a** pH in groundwater. **b** pH in surface water

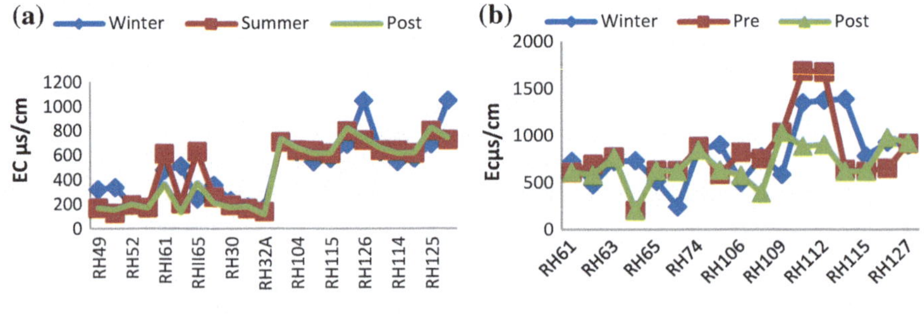

Fig. 4 **a** EC in groundwater. **b** pH in surface water

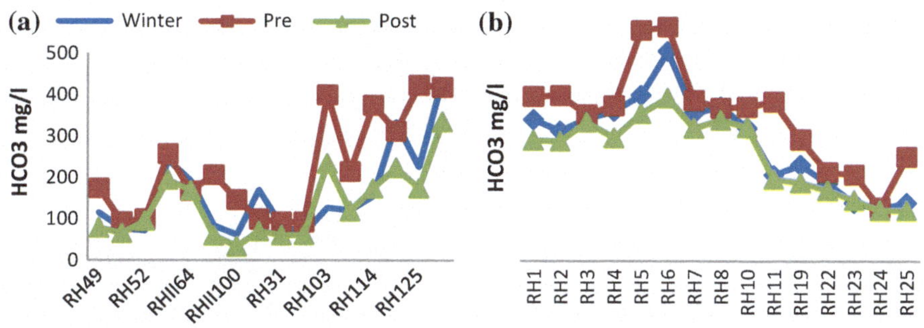

Fig. 5 **a** Bicarbonate in groundwater. **b** Bicarbonate in surface water

3.1.5 Nitrate

According to BIS standard (2012) (BIS (IS 1050) for drinking water quality, the desirable limit for nitrate is 45 mg/l and maximum permissible limit is 100 mg/l, for drinking water beyond this a disease called methemoglobinemia occur. WHO has given the most desirable limit for NO_3 as 50 mg/l for drinking water. Higher concentration of nitrate (>45 mg/l) was observed at Rispana River (84 mg/l), the confluence point of Bindal, Suswa and Rispana River), Bindal River at Nalapani and in handpump water of Patel Nager, Dehradun (Figs. 6 and 7).

P does not exist as the free element in nature but is widely distributed in rock minerals, plants and animals. It is the 12th most abundant element in the earth crust and found approximately about 0.12%. Although phosphate was not detected in study area at number of locations in all the three seasons but in few sampling locations the phosphate in water was above the permissible limit of WHO. The higher PO_4 concentration that is above WHO drinking water limit had been found in Bindal River at Bhamanwala (18 mg/l), Suswa, Rispana and Bindal River in their lower reaches. The presence of phosphate and nitrate at these locations indicate adverse impact of urbanization on water quality.

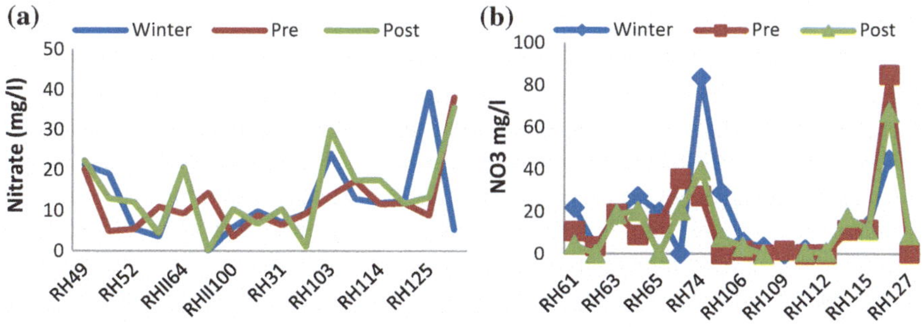

Fig. 6 **a** Nitrate in groundwater. **b** Nitrate in surface water

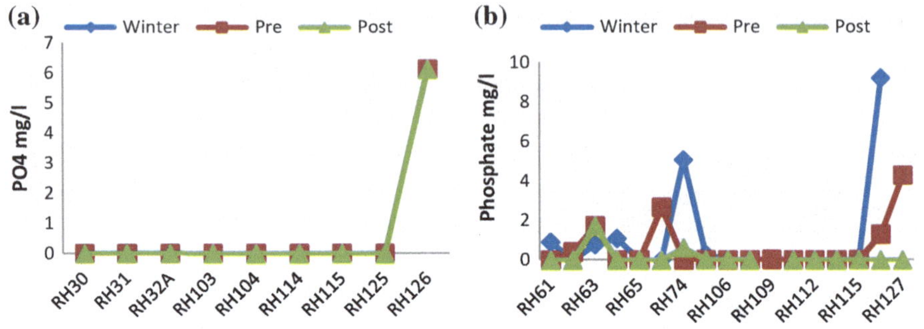

Fig. 7 **a** PO_4 in groundwater. **b** PO_4 in surface water

4 Heavy Metals

4.1 Nickel

The Ni concentration in Winter season varied from 0.001 ppb to 0.396 ppb with an average of 0.396 ppb, during Summer season it varies from 0 to 0.38 ppb with an average of 0.396 ppb and in post-monsoon it varies from 0 ppb to 0.39 ppb with an average of 0.39 ppb in the entire study area. In Dehradun industrial area of groundwater of Doon, the Ni concentration found to be almost same ~0.46 in all three seasons. In agricultural area of Dehradun groundwater the Ni concentration did not show much variation and found almost ~0.54 ppb in all three seasons.

In urban area of Dehradun groundwater the average Ni concentration is ~0.24 in all three seasons. In surface water of Dehradun, the nickel concentration did not show much variation and found to be almost same ~0.4 ppb in Summer, post-monsoon and in Winter season.

4.2 Lead

The concentration of lead in all sampling location (5 samples) were found to be less than 1 ppb. The Ni concentration in Winter season varied from 0.001 ppb to 0.396 ppb with an average of 0.396 ppb, during Summer season it varied from 0.001 to 0.38 ppb with an average of 0.396 ppb and in post-monsoon it varies from 0.001 ppb to 0.39 ppb with an average of 0.39 ppb in the entire study area. In Dehradun industrial area of groundwater of Doon, the Ni concentration found to be almost same ~0.46 in all three seasons. In agricultural area of Dehradun groundwater the Ni concentration did not show much variation and found almost ~0.54 ppb in all three seasons.

4.3 Cd

The Cd concentration in industrial, agricultural, urban of groundwater of Dehradun and districts and in surface the amount of Cd at all sampling location was found to be under below detectable limit.

4.4 Cr

In industrial area of groundwater the Cr concentration in few sampling locations was found high ~5889 ppb during Winter season, Summer season and post-monsoon season and at Selaqui Industrial Area (Cr 5990 ppb) during Summer season and Selaqui Industrial Area (1.2 ppb) during Winter season and during Summer it was 1.3 ppb while in rest of the sampling location it was below the detectable limit. In urban area of Doon, the Cr was found only at the sampling location Yamuna River at Vikasnager, during Winter season Cr concentration was 0.08 ppb and during post-monsoon season its concentration was 0.084 ppb.

In surface water of Doon, the Cr was found at the sampling location of Rispana River (Winter season 16.9 ppb and Summer season 17.8 ppb) and at Suswa River (Winter season—987 ppb) and during (Summer 1.97 ppb) whereas in rest of the sampling locations it was found below the detectable limit. Except for two sampling locations, i.e. Selaqui Industrial area and Suswa River the Cr concentrations in all sampling location found to be under maximum permissible limit 50 ppb of BIS and WHO standards for drinking water quality. The possible sources of Cr in groundwater of industrial area is leaching of effluents that may contain toxic metals in surface water of Suswa River, it may come from urban waste that may contain (E-waste, i.e. batteries, cells and other waste of electronic devices).

4.5 Cu

In Dehradun industrial area the copper concentration was 25 ppb, 24.7 ppb and 0.6 ppb, respectively, in Summer, post-monsoon and in Winter season. In agricultural area of Dehradun groundwater the average concentration of the Cu was 4.08 ppb, 13.4 ppb and 5.33 ppb, respectively, in Summer, post-monsoon and in Winter season. In urban area of Dehradun groundwater the Cu concentration did not show much variation and found to be almost same ~20.7 in all three seasons. In surface water Dehradun also the Cu concentration did not show much variation and found to be almost same ~0.4 ppb in all three seasons.

The high concentrations of Cu were found at confluence point of Bindal, Rispana and Suswa River and crossed the desirable limit of Cu (50 ppb) of BIS standard of drinking water quality but found to be under maximum permissible limit of BIS (1500 ppb) at few sampling locations.

4.6 Mn

The concentration of Mn varies from < 1 ppb to 142 ppb, the highest concentration of Mn was found at few the sampling location of Bharuwala Cant area, shallow handpump, confluence point of Bindal, Rispana and Suswa River. One sample at Bahruwala (142 ppb) had crossed the maximum permissible limit of WHO standards of drinking water quality. According to BIS standard of drinking water quality few samples crossed the maximum desirable limit (100 ppb) but found under the maximum permissible limit (300 ppb) of BIS for drinking water quality.

4.7 Co and Zn

The concentration of cobalt found to be less than one ppb and high concentration of Co is found at the sampling location Bharuwala Cantt area (1 ppb). The Zn concentration varies from 17 ppb to 170 ppb in study area. All the samples were under permissible limit (15000 ppb) of BIS standards for drinking water quality.

4.8 Arsenic and Fe

The As concentrations varied from 2 ppb to 8 ppb in study area and found to be under the permissible limit (50 ppb) of BIS standards of drinking water quality. The Fe concentrations in study area varied from 37 ppb to 228 ppb in all sampling location.

4.9 Boron and Sr

The B concentration was ranged from 335 ppb to 8 ppb in surface and groundwater. The B concentration is found maximum at sampling location at Rispana River (335 ppb).

The concentration of **Strontium** (Sr) was found high in all sampling locations Rispana River (104 ppb), Bharuwala Cantt area (371 ppb) and Selaqui industrial area (262 ppb) (Table 1).

Table 1 Trace metals (ppb) in Winter season in Dehradun and comparison with BIS standard IS 10500: 2012

Range	Pb	Mn	Cu	Zn	Fe	Co	As	Sr	Ni	Cr	Si	Cd	B
Min	<1	4	1	16	37	1	0	75	1.17	5	1.5	<1	8
Max	<1	142	290	170	236	1	8	371	9	31	7.3	<1	335
Average	<1	54	18.9	68.4	153.8	1	3	236	4.25	15.8	4.26	<1	85.4
BIS (Required acceptable limit)	10	100	50	5000	300		10		20	50			500
(Permissible limit in absence of alternate source)	No relaxation	300	150	15000	No relaxation		15		No relaxation	No relaxation			1000

5 Conclusion

The assessment of drinking water quality of the study area was carried out by comparing the chemical data with BIS standard for drinking water quality 2012. Majority of the chemical parameter falls under the permissible limit of drinking water but few parameters, i.e. F, NO_3 and SO_4. crossed maximum permissible limit of BIS standards of drinking water quality. All heavy metal parameter, i.e. Pb, Mn, Zn, Fe, As, Ni, Cr and B were found under the permissible limit of BIS standard (IS 10500: 2012) of drinking water quality except Cu which has crossed the maximum permissible limit of BIS standard at few sampling locations.

References

Abbas N, Subramanain V (1984) Erosion and sediment in Ganga river basin (India) J Hydrol 69:3–182

APHA (1992) Standard methods for examination of water and waste waters, 18th edn. American Public Health Association, Washington, DC

Bahukhandi DK, Aaron A (2017) 2016/JOGEE/4499 impact of improper disposal of municipal solid waste on ground water quality in and around the solid waste dumping sites of Visakhapatnam, Andrapradesh, Indian. J Glob Ecol Environ Int Knowl Press 5(3):133–143. 27 Feb, 2017. ISSN: 2454-2644, http://www.ikpress.org/issue/836

Bahukhandi DK, Bartarya SK (2012) "Hydrochemistry of surface and groundwater of Doon valley" has been presented and published in proceeding volume on fifth international groundwater conference (IGWCC) jointly organized by association of geologist and hydrologist at Aurangabad, Maharashtra, vol. 3, Paper No- 74, pp 1109–1119

Bahukhandi KD, Bartarya SK (2014) Major ion chemistry of Asan River catchment of Dehradun district (impact assessment of anthropogenic sources). Octa J Environ Res 2(2):168–177. International Peer-Reviewed Journal ISSN 2321 3655. http://www.sciencebeingjournal.com (Impact Factor: 0.64) Science Being Publisher

Bahukhandi DK, Siddiqui NA (2008) Impact assessment of Urbanization on water quality. Environ Pollut Control J 12(1)

Bahukhandi KD, Bartarya SK, Siddiqui NA(2017) Assessment of surface and ground water quality of Haridwar district of Uttarakhand. Int J Chem Tech Res. 10(10), 95–118. http://www.sphinxsai.com/2017/ch_vol10_no10/1/(95-118)V10N10CT.pdf). ISSN: 0974-4290, ISSN(Online):2455-9555 (Scopus Index)

Bahukhandi KD, Satendra, Siddiqui NA, Singh R, Arora S, Mondal P (2018) Assessment of major ion chemistry and spatial variation of water quality of ganga river of Uttarakhand. J Chem Pharm Res 10(7):145–150. ISSN: 0975-7384

Bartarya SK, Bahukhandi DK (2012) "Impact assessment of urbanization and industrialization on surface and groundwater quality of Dehradun district of Uttarakhand. Global J Eng Design Techn (Pub: Global Institute of Research & Education), GJET **1**(1):11–22. ISSN 2319-7293

Bhatt KP, Saklani S (1996) Hydro geochemistry of the upper Ganges river India. Jornel Geol Soc India 48:171–182

BIS (IS 10500: 2012) (2012) Standard of Drinking Water Quality

Chakrapani GJ (2002) Water and sediment geochemistry of major Kumaun Himalayan lakes India. Environ Geol 2002(43):99–107

Chakrapani GJ (2005) Major & trace element geochemistry in upper Ganga river in the Himalaya India. Environ Geol 48:189–201

Chakrapani GJ, Subramanian V (1996) Fractionation of heavy metals and phosphorus in suspended sediments of the Yamuna river India. Environ Monit Assess 43:117–124

Chakrapani GJ, Subramanian V, Gibbs RJ, Jha PK (1995) Size characteristic and mineralogy of suspended sediments of the Ganges river India. Environ Geol 25:192–196

Dalai TK, Krishnaswami S, Sarin MM (2002a) Major ion chemistry in the in the headwater of the yamuna river system: chemical weathering, its temperature dependence and CO2 consumption in the Himalaya. Geochimica et Cosmichimica Acta 66(19): 3397–3416

Dalai TK, Krishnaswami S, Sarin MM (2002b) Barium in the yamuna river system in the himalaya: sources, fluxes and its behavior during weathering and transport. Geochem Geophys Geosyst 3. https://doi.org/10.1029/2002gc000381

Galy A, France-Lanord C (1999) Weathering processes in the Ganges-Brahmputra basin and the riverine alkalinity budget. Chem Geol 159:31–60

Gibbs RJ (1970) Mechanism controlling world water chemistry. Science 170:1088

Kanchan B, Prasanjeet M, Satendra S (2015) Water quality assessment of river ganga health hazard identification and control singh. Int J Sci Res Publ 5(5). ISSN 2250-3153, www.ijsrp.org (impact factor: 3.27)

Krishnaswami S, Trivedi JR, Sarin MM, Ramesh R, Sharma KK (1992) Strontium isotope and rubidium in the Ganga-Brahmputra river system: Weathering in the Himalaya, fluxes to the Bay of Bengal and contribution to the evolution of oceanic 87Sr/86Sr. Earth Planet Sci Lett 109:243–253

Livingstone DA (1963) The sodium cycle and the age of ocean. Geochim Cosmochim Acta 27:1055

Meybeck M (1976) Total mineral dissolved transported by world major rivers. Hydro Sci Bull 21:265

Pande K, Sarin MM, Trivedi JR, Krishnaswami S, Sharma KK (1994) the Indus river system (India-Pakistan): Major ion chemistry, uranium and strontium isotopes. Chem Geol 116:245–259

Pandey DK, Rai JPN, Kumar O (1991) Studies on water quality of fish pond water at Pant Nagar. Indian. J Environ Prot 13(6): 451–453

Raymahasay BS (1970) Characteristics of stream erosion in the Himalayan region of India. In: Proceeding of the symposium in hydro geochemistry and biogeochemistry, vol 1, pp 82–89

Subudhi RK, Chakrapani GJ, Saini RK (2007) Temporal variation in dissolved elemental composition in the Ganga river at Risikesh, Uttaranchal. J Geol Soc India 69

Hydrological Assessment of River Henwal Using Water Quality Indices with Reference to Planktonic Composition

Gagan Matta, Xi Xi Lu and Avinash Kumar

Abstract Henwal is a spring-fed river and a source of freshwater in Garhwal region, India. It is one of the important tributaries of River Ganga and its water is useful for irrigation and hydropower production. The purpose of this study was to check the pollution level in river water by using the comprehensive pollution index (CPI). The assessment of CPI was based on different hydrological parameters. CPI was found within the range of 2.11–4.66 which indicates that the river water was severely polluted. The diversity and density of different planktonic groups were also observed and found to be higher in at Jajal site. Blue-green algae dominate phytoplanktons, and *Protozoa* was dominant among zooplankton. It is recommended that pollution control practices should be adopted and planktonic composition should be monitored on a regular basis.

Keywords River Henwal · River Ganga tributary · CPI · Planktons · Uttarakhand · Chamba · Simpson and Shannon–Weaver index

1 Introduction

Freshwater is one of the key sources of survival of a living form on earth. Freshwater is the water with less than 0.5 per thousands of dissolved salts. On earth, the freshwater systems consist of rivers, streams, lakes, ponds, groundwater, cave water, springs, floodplains, wetlands, etc., completing the requirement of all the elements of nature available (Matta 2010). For humans, freshwater provides water from drinking to sanitation, agriculture, industry, transport, electricity generation and recreation, and many more (Matta and Kumar 2017a; Matta et al. 2017). For animals and plant species, these freshwater bodies are a source for drinking and also their habitat. River systems are the nervous system of the planet, pumping freshwater into wetlands and

---

G. Matta (✉) · A. Kumar
Department of Zoology & Environmental Science, Gurukula Kangri University, Haridwar, India
e-mail: drgaganmatta@gkv.ac.in

X. X. Lu
Department of Geography, National University of Singapore, Singapore, Singapore

N. A. Siddiqui et al. (eds.), *Advances in Water Pollution Monitoring and Control*,
Springer Transactions in Civil and Environmental Engineering,
https://doi.org/10.1007/978-981-32-9956-6_18

lakes and out to sea. They are one of the major sources of nutrients flushing through aquatic ecosystems, helping the ecological cycle and ecosystems to sustain while keeping thousands of species alive, helping the fisheries industries of worth billions of dollars. For human civilization, freshwater is the major component for cities, farms, and factories (Matta et al. 2018a, b, c, d, e).

India with approximately 4% of world's freshwater, ranked among top ten water-rich countries, but according to the Working Group II report of the Intergovernmental Panel on Climate Change, Fourth Assessment designated India as "water stressed region". Water demand is on a high with the rapid increase in population causing rapid urbanization and industrialization with modern demand for agriculture. In India, 14 major river systems share about 83% of the drainage basin (Uttarakhand Tourism Development Board 2018; Census 2011). The Himalayas, the source of fresh water rivers, supports the life of many ecosystems in India, receives about 20–30% of their water from melted snow during the summer and monsoon seasons (Matta et al. 2018d).

River Henwal is spring-fed river, a tributary to River Ganga in the Garhwal region, Uttarakhand state of India, lying between the latitude of 29°26′–31°28′ N and longitude 77°99′–80°6′ E with the covered area of 30,000 km (approx.). The origin is Kaddukhal (Near Surkanda Devi Temple, Chamba, Uttaranchal) nearly about 38 km. The river is divided into three different ecological zones, viz., the upper, the middle, and the lower section. The substratum of the river is rocky, stony, pebbly, and finally, sandy when it meets River Ganga. With the fluctuation in velocity and temperature of the water, the ecological systems may be different. The present study was conducted to assess and validate the influence of anthropogenic activities on water quality in terms of ecological characters of River Henwal, a tributary of River Ganga. The present study has been carried out by considering the importance of aquatic ecosystems of the Himalayas with special reference to Garhwal region (Matta and Kumar 2017b).

2 Materials and Method

2.1 Study Area

The River Henwal basin is located between latitude 30°20′43.34″ N and 78°23′40.83″ E, in the Chamba Dist. Tehri Garhwal, at an elevation of 1600 m above sea level, is located in the newly carved Uttarakhand state. The River Henwal is one of the important and spring-fed tributaries of the Ganga river that originates from Kaddukhal (near Surkanda Devi Temple, Chamba, Uttarakhand) covering the catchment area of ~250 km^2 with the length of about 38 km. The river substratum is stony, rocky, pebbly, and finally, sandy when it meets River Ganga. With the significant variation in the velocity and temperature of the water, the biodiversity may also vary (Matta et al. 2017, 2018a). The regional climatic condition of the study area lies within the range of about 15–30 °C during summer and −5 to 20 °C during winter (Uttarakhand Tourism Development Board 2018). The population of Chamba is 7,771 (Census 2011), surrounded by famous tourist spots like Mussoorie, Rishikesh, and Tehri Dam reservoir, New Tehri, connected by a junction of roads. The study area is a small hill station, 50 km far from Mussoorie, which is till now an unknown destination to most tourists, because of which that area is unpolluted and fresh to breathe for a few lucky people—those looking to experience adventure and natural beauty. There is a richness of pine and deodar trees which give a good and pleasant view (Matta et al. 2018d).

The river water is used for irrigation, domestic, and other purposes. During the study, hydrochemical and biological parameters of the river water were analyzed and assessed to represent the quality of River Henwal. Samples were taken from three sites of the Henwal river, i.e., control site (upstream and undisturbed site), Nagni, Khadi (midstream and least disturbed site), and Jajal (downstream and highly disturbed site) (Fig. 1).

2.2 Sampling and Analysis Procedure

The sampling for water quality assessment was carried out for four different seasons, i.e., winter (December to February), summer (March to May), monsoon (June to August) and post-monsoon (September to November) in the study area during 2015–2016. The collection of water samples was done in acid-washed Nalgene-Wide-Mouth Natural HDPE polypropylene bottles (1000 mL). Before the collection of samples, the bottles were thoroughly rinsed with river water and air bubble were avoided during bottle fill-up. In situ analysis of some physico-chemical parameters (temperature, conductivity, and dissolved oxygen: DO) were done with the help of a Portable Multi-Parameter Instrument, Model—TMULTI 27 (TOSHCON) (Matta and Gjyli 2016). The sealed water bottles were immediately transferred to the laboratory for the observation of other remaining physico-chemical parameters as per the prescribed standard procedure (Matta and Uniyal 2017).

2.3 Statistical Analysis

In the statistical analysis, different indices like CPI, Simpson, and Shannon–Wiener were calculation and presented graphically by the use of Minitab 16.

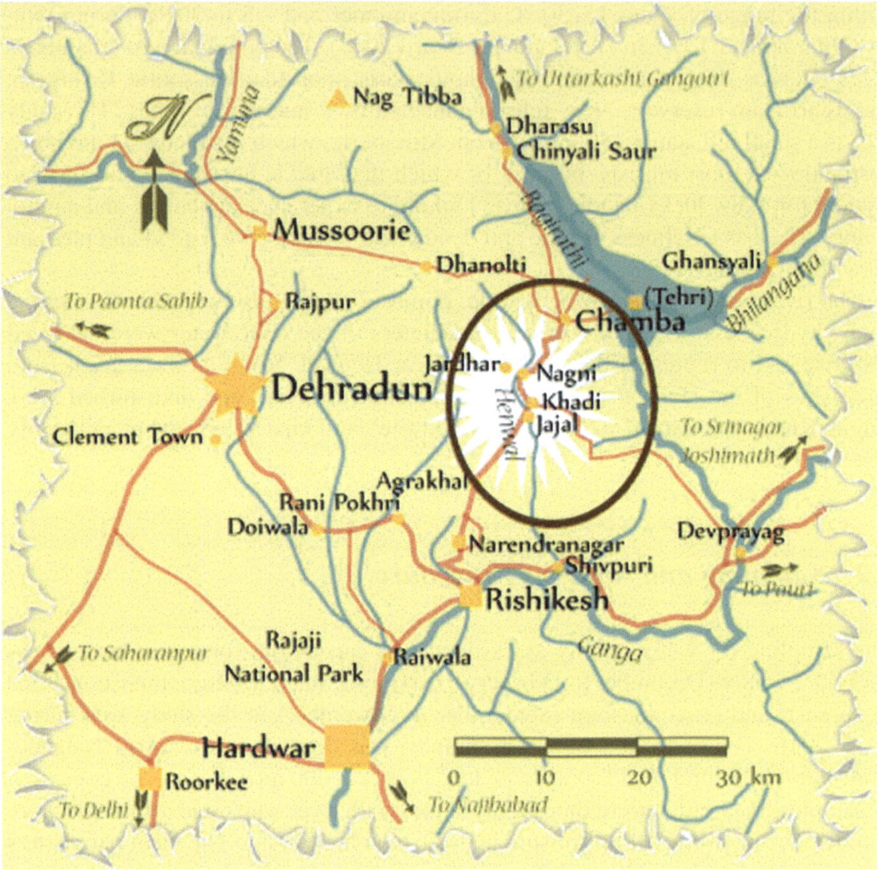

Fig. 1 Showing sampling locations

2.4 *Comprehensive Pollution Index (CPI)*

The calculation and classification of CPI to classify the water quality status has been followed by many of the researchers (Matta 2014; Matta and Kumar 2015; Gautam 2017; American Public Health Association (APHA) 2012). The evaluation of CPI was by the following equations:

$$\mathrm{PI} = \frac{\text{Measured concentration of invididual parameter}}{\text{Standard permissible concentration of parameter}} \tag{1}$$

$$\mathrm{CPI} = \frac{1}{\mathrm{n}} \sum_{\mathrm{i}=0}^{\mathrm{n}} \mathrm{PI} \tag{2}$$

where PI represents the pollution index of separate water quality parameter measured, and CPI is the comprehensive pollution index. The standard permissible concentrations of each parameter considered in the study were proposed by different government agencies for a general discharge of environmental pollutants (Matta et al. 2018a, b, c, d, e; Zhao et al. 2012; Mishra et al. 2015). The rating scale of CPI ranges from 0 to 2 which are as follows: ≤ 0.20 is excellent; 0.21–0.40 is good; 0.41–1.00 is fair; 1.01–2.0 is poor; and ≥ 2.01 is very poor quality.

2.5 *Simpson's Diversity Index (D)*

Simpson's diversity index (D) is calculated using the following equation (Environmental Protection Agency (EPA) 2012):

$$D = 1 \Big/ \sum (P(i))^2$$

where p(i) represents the portion of the overall sample comprised by the species,

$$\mathrm{p(i) = n(i)/N}$$

where n(i) is the individual's numbers and N is the total number of all species.

2.6 *Shannon–Weaver Index (H)*

This index method is used for calculating biotic diversity in water bodies and terrestrial ecosystems. The calculation of H is given as follows (Bureau of Indian Standards (BIS) 2012):

$$\mathrm{H} = -\sum \mathrm{P(i)\ln\ P(i)}$$

where p(i) represents the portion of the overall sample comprised by the species,

$$\mathrm{p(i) = n(i)} \Big/ \mathrm{N}$$

where n(i) is the individual's number and N is the total number of all species.

2.7 Taxon Evenness

Taxon evenness is the relative distribution of individuals among taxonomic groups within a community) and is expressed as

$$E = H/\ln(R)$$

where E indicates the Taxon Evenness and R represents the Taxon Richness defined as the total no. of distinct taxa in a population.

3 Result and Discussion

The hydrochemical compositions of the river water samples were analyzed and the site-wise observed dataset were presented seasonally in Table 1. On the basis of this analysis, the higher velocity (1.12 ± 0.12 m/s) of the river was recorded at Jajal during autumn whereas the lower velocity (0.60 ± 0.00 m/s) was recorded at Nagni during summer. The high velocity (2.18 m/s) in monsoon and low (0.39 m/s) in winters were recorder for River Ganga at Haridwar (World Health Organization (WHO) 2011). The low velocity of a river has less potential to disperse the pollutants (Mishra et al. 2015). Water temperature was noticed within the range 13–28 °C throughout the study period. The maximum conductivity (129.46 S/cm) and minimum (96.78 ± 0.02 S/cm) was found at the control site.

The presence of organic pollutants, runoff, and rainfall makes the river water quality more turbid. The turbidity of the Ganga river water was recorded more than the permissible limit of Indian Standards (IS) and WHO during the monsoon season (Simpson 1949). The maximum turbidity (349.15 ± 25.00 JTU) was noticed at Jajal during autumn and minimum (16.79 ± 1.91 JTU) at Nagni during winters. Maximum level (534.71 ± 2.40 mg/l) of total solids was recorded at Khaadi during monsoon and minimum (54.78 ± 13.71) was noticed at Nagni during winters. Similar findings were observed from the Baldi stream of Garhwal Himalayas (Shannon and Weaver 1949). The higher level (10.3 mg/l) of dissolved oxygen (DO) was observed at the control site during winters whereas the lowest level (7.4 mg/l) was recorded at Jajal during summer. Some observations of different researchers (Joshi et al. 2009; Matta et al. 2015a, b) supported the same findings of DO for River Ganga. Even though the river water has sufficient DO levels (above 6 mg/l) as compared to the standard limit proposed by WHO and BIS, the high photosynthetic activity of phytoplanktons could be the reason for the appropriate DO (Sharma et al. 2016; Matta 2015). Similiar observations were recorded for many rivers of the north Indian River plain (Chauhan and Singh 2010). On the contrary of dissolved oxygen, higher concentration (1.8 ± 0.90 mg/l) of free CO_2 was observed at Jajal in summer and lower concentration (0.2 ± 0.00 mg/l) at the control site during winters. Similar observations were reported

Table 1 Seasonal variation in the mean concentration of different hydrological parameters of different sites

Parameters	Sites with season							
	Control site				Nagni			
	Autumn	Winter	Summer	Monsoon	Autumn	Winter	Summer	Monsoon
Velocity	1.04 ± 2.15	0.78 ± 0.01	0.76 ± 1.18	0.98 ± 1.62	0.75 ± 0.02	0.63 ± 0.02	0.60 ± 0.00	0.66 ± 0.05
Temperature	19.4 ± 0.03	15.0 ± 1.74	27.0 ± 2.14	23.0 ± 0.27	13.0 ± 0.25	15.0 ± 0.04	20.0 ± 1.07	24.0 ± 1.06
Conductivity	129.46 ± 0.00	96.78 ± 0.02	123.10 ± 0.21	119.50 ± 0.15	99.79 ± 0.42	99.63 ± 0.00	109.19 ± 0.02	115.17 ± 0.04
Turbidity	242.10 ± 2.10	59.41 ± 2.10	209.70 ± 37.86	279.86 ± 120.00	55.34 ± 1.02	16.79 ± 1.91	30.25 ± 0.05	56.12 ± 2.15
Total solids	354.74 ± 11.60	82.25 ± 0.25	239.79 ± 0.16	350.75 ± 0.24	101.24 ± 0.85	54.78 ± 13.71	102.51 ± 0.19	124.41 ± 2.15
DO	9.6 ± 0.20	10.3 ± 0.00	7.9 ± 0.00	8.5 ± 0.05	10.3 ± 0.32	8.9 ± 0.57	8.8 ± 0.02	7.8 ± 1.03
Free CO_2	0.6 ± 0.04	0.2 ± 0.00	0.4 ± 0.12	0.7 ± 0.00	0.6 ± 0.00	0.2 ± 0.00	0.6 ± 0.01	1.3 ± 0.05
Acidity	37.56 ± 0.05	52.76 ± 1.24	49.32 ± 0.07	45.54 ± 0.25	49.85 ± 0.03	59.21 ± 1.16	44.37 ± 0.16	39.92 ± 0.26
Total Alkalinity	201.5 ± 0.05	204.0 ± 5.64	208.0 ± 0.06	172.0 ± 0.04	198.0 ± 35.01	165.0 ± 0.76	159.4 ± 0.06	148.5 ± 1.03
Total Hardness	167.0 ± 0.02	148.0 ± 1.65	213.0 ± 0.40	173.0 ± 3.08	184.0 ± 0.06	130.0 ± 0.05	115.0 ± 0.10	142.0 ± 0.02
Ca Hardness	46.70 ± 2.30	35.92 ± 0.21	23.47 ± 0.88	32.50 ± 1.62	32.17 ± 0.26	32.74 ± 3.01	28.78 ± 0.03	23.50 ± 0.00
Mg Hardness	18.95 ± 0.19	18.40 ± 2.38	27.82 ± 0.59	22.63 ± 0.81	26.30 ± 0.10	16.86 ± 5.26	13.87 ± 0.24	15.96 ± 0.05
Chlorides	8.28 ± 0.50	8.96 ± 7.10	6.64 ± 0.16	6.12 ± 0.25	7.29 ± 0.05	7.99 ± 0.09	8.25 ± 0.08	7.97 ± 0.02
Phosphates	0.34 ± 0.11	0.55 ± 0.10	0.79 ± 0.09	0.43 ± 0.15	0.61 ± 0.11	0.38 ± 0.18	0.38 ± 0.17	0.56 ± 0.21
Zn	0.18 ± 0.25	0.13 ± 0.21	0.14 ± 0.17	0.18 ± 0.23	1.46 ± 0.32	1.87 ± 0.79	1.61 ± 0.41	0.16 ± 0.2
Pb	0.03 ± 0.04	0.03 ± 0.04	0.03 ± 0.04	0.14 ± 0.19	0.86 ± 0.04	0.74 ± 0.05	0.99 ± 0.05	0.2 ± 0.28
Mn	0.16 ± 0.27	0.18 ± 0.3	0.15 ± 0.25	0.18 ± 0.3	0.45 ± 0.27	0.45 ± 0.18	0.41 ± 0.12	0.17 ± 0.29

(continued)

Table 1 (continued)

Parameters	Sites with season							
	Control site				Nagni			
	Autumn	Winter	Summer	Monsoon	Autumn	Winter	Summer	Monsoon
Fe	0.02 ± 0.02	0.02 ± 0.02	0.01 ± 0.02	0.02 ± 0.02	0.05 ± 0.04	0.03 ± 0.01	0.03 ± 0.01	0.02 ± 0.02
Cu	0.19 ± 0.32	0.15 ± 0.2	0.11 ± 0.14	0.14 ± 0.19	0.29 ± 0.04	0.39 ± 0.35	0.67 ± 0.4	0.17 ± 0.23
Al	0.02 ± 0.03	0.03 ± 0.04	0.03 ± 0.04	0.03 ± 0.04	2.12 ± 0.85	0.28 ± 0.26	0.76 ± 0.1	0.3 ± 0.51
Cd	0.19 ± 0.32	0.21 ± 0.36	0.2 ± 0.33	0.17 ± 0.29	0.27 ± 0.19	0.3 ± 0.2	0.26 ± 0.24	0.2 ± 0.33
Mg	0.02 ± 0.02	0.01 ± 0.02	0.02 ± 0.03	0.02 ± 0.02	0.14 ± 0.19	0.18 ± 0.26	0.16 ± 0.24	0.02 ± 0.02
Velocity	1.02 ± 0.73	0.71 ± 1.24	0.84 ± 1.25	1.01 ± 0.75	1.12 ± 0.12	0.89 ± 0.00	0.68 ± 0.02	1.07 ± 0.05
Temperature	19.0 ± 1.21	15.0 ± 0.04	28.0 ± 0.57	24.0 ± 1.18	20.5 ± 0.50	16.0 ± 0.00	27.0 ± 0.26	23.50 ± 0.60
Conductivity	128.47 ± 0.00	99.73 ± 0.00	120.26 ± 0.04	125.46 ± 0.16	123.75 ± 0.03	112.91 ± 0.05	123.78 ± 0.03	122.78 ± 0.03
Turbidity	203.45 ± 22.07	31.93 ± 0.16	226.43 ± 32.52	279.71 ± 127.86	349.15 ± 25.00	128.95 ± 2.80	180.46 ± 62.00	329.89 ± 3.86
Total solids	375.24 ± 0.50	81.71 ± 0.06	211.67 ± 0.16	534.71 ± 2.40	498.67 ± 46.00	215.75 ± 72.50	186.41 ± 0.16	442.64 ± 0.06
DO	9.4 ± 0.53	9.4 ± 0.81	8.3 ± 0.24	8.7 ± 0.16	8.9 ± 1.16	9.7 ± 0.23	7.4 ± 0.12	8.3 ± 0.09
Free CO_2	0.4 ± 0.00	0.3 ± 0.03	0.5 ± 0.02	0.5 ± 0.00	0.7 ± 0.00	0.6 ± 0.01	1.8 ± 0.90	1.0 ± 0.10
Acidity	36.52 ± 7.06	53.99 ± 0.05	53.67 ± 0.08	45.15 ± 0.05	42.12 ± 0.30	31.84 ± 0.25	48.90 ± 0.45	28.25 ± 0.17
Total Alkalinity	164.0 ± 0.27	183.0 ± 1.65	205.0 ± 0.12	182.0 ± 0.26	121.0 ± 0.14	181.0 ± 7.39	223.0 ± 5.68	137.0 ± 0.01
Total Hardness	154.0 ± 0.21	161.0 ± 0.02	230.0 ± 0.41	198.0 ± 0.12	163.0 ± 12.56	192.0 ± 0.83	246.0 ± 0.11	189.0 ± 0.48
Ca Hardness	34.23 ± 0.06	33.10 ± 2.14	24.96 ± 0.52	27.89 ± 0.24	31.50 ± 1.60	42.40 ± 1.15	27.43 ± 0.34	35.14 ± 0.12

(continued)

Table 1 (continued)

Parameters	Sites with season							
	Control site				Nagni			
	Autumn	Winter	Summer	Monsoon	Autumn	Winter	Summer	Monsoon
Mg Hardness	19.12 ± 0.11	19.85 ± 0.24	28.37 ± 0.25	21.10 ± 0.15	20.96 ± 0.48	18.34 ± 3.34	24.60 ± 0.00	21.00 ± 1.27
Chlorides	12.45 ± 0.24	8.64 ± 0.03	6.98 ± 0.36	6.21 ± 0.00	7.59 ± 6.08	8.56 ± 0.70	6.30 ± 0.01	7.82 ± 0.65
Phosphates	0.28 ± 0.12	0.58 ± 0.18	0.86 ± 0.18	0.45 ± 0.20	0.35 ± 0.09	0.41 ± 0.10	0.82 ± 0.17	0.46 ± 0.06
Zn	1.54 ± 0.27	1.87 ± 0.38	1.84 ± 0.82	1.56 ± 0.39	1.19 ± 0.14	0.63 ± 0.54	1.71 ± 0.75	2.25 ± 1.07
Pb	0.85 ± 0.03	1.09 ± 0.27	0.94 ± 0.02	0.92 ± 0.03	0.76 ± 0.13	1.48 ± 0.73	0.69 ± 0.25	1.17 ± 0.4
Mn	0.41 ± 0.07	0.52 ± 0.15	0.55 ± 0.35	0.56 ± 0.34	0.67 ± 0.23	0.32 ± 0.21	0.41 ± 0.19	0.4 ± 0.21
Fe	0.03 ± 0.01	0.03 ± 0.01	0.03 ± 0.01	0.03 ± 0.01	0.03 ± 0.01	0.05 ± 0.02	0.04 ± 0.01	0.05 ± 0.02
Cu	1.61 ± 0.57	1.5 ± 0.59	1.24 ± 0.23	1.93 ± 0.42	1.84 ± 0.53	1.59 ± 0.46	1.86 ± 0.35	1.78 ± 0.41
Al	0.91 ± 0.14	0.83 ± 0.22	1.06 ± 0.21	0.94 ± 0.07	1.01 ± 0.17	0.94 ± 0.16	1.07 ± 0.16	0.83 ± 0.14
Cd	0.29 ± 0.05	0.32 ± 0.06	0.31 ± 0.09	0.32 ± 0.02	0.29 ± 0.09	0.32 ± 0.1	0.32 ± 0.06	0.3 ± 0.11
Mg	0.02 ± 0.01	0.03 ± 0.01	0.03 ± 0.01	0.03 ± 0.01	0.03 ± 0.01	0.03 ± 0.01	0.03 ± 0.01	0.03 ± 0.01

for CO_2 (Sharma and Rathore 2000). It is simply the respiration process of plankton responsible for changing productivity level.

The higher level (59.21 ± 1.16 mg/l) of acidity was observed at Nagni whereas the lower level (31.84 ± 0.25 mg/l) was reported at Jajal during winters. The alkalinity observation exhibited ahuge variation throughout the study period. The high alkalinity (223.0 ± 5.68 mg/l) and low alkalinity (121.0 ± 0.14 mg/l) was reported at Jajal in summer. The highest level of hardness (246.0 ± 0.11 mg/l) was observed at Jajal and lowest (115.0 ± 0.10 mg/l) was at Nagni during summer. The acceptable limit of the total hardness for drinking water is 200 mg/l, according to BIS 2012. Many of the observations revealed that the major source of hardness ions is domestic wastewater (Ravindra et al. 2003; Rani et al. 2011; Singh 2014). The Ca–Mg harness was observed under the permissible limit of BIS 2012 (75 and 30 mg/l, respectively). The chloride concentration was varied from 6–12 mg/l and found to be under the desirable limit (250 mg/l) proposed by BIS. The various sources of chlorides in surface water are agricultural runoff, rocks containing chlorides, oil well wastes, wastewater treatment plant effluents, and industrial wastewater. A very less concentration of chloride is vital for normal cell function whereas, in high levels of chlorides, fish and other aquatic living organisms cannot survive (Haritash et al. 2016). Nutrients like phosphate compounds influence the life form of any aquatic body in terms of growth and photosynthetic activity and lead to increase levels of eutrophication (Badrakh et al. 2008). The higher level (0.86 ± 0.18 mg/l) of phosphate was evaluated during summer and the lower level (0.28 ± 0.12 mg/l) was analyzed in winters at Khaadi. A similar observation was recorded for the Ganga river water at Rishikesh (Uttarakhand) India (Matta et al. 2017).

Heavy metal pollution is the cause of most kinds of diseases occurring to the exposed organisms. This study concludes the assessment of eight heavy metals (Zn, Pb, Mn, Fe, Cu, Al, Cd, and Mg) of River Henwal. Only the concentrations of Zn, Fe, and Mg were recorded under the proposed desirable limit of BIS for drinking water, rest of the metals (Pb, Mn, Cu, Al, and Cd) were found above the standard limit during all seasons at all sites. The higher concentrations of Pb, Mn, Cu, Al, and Cd was 1.48 ± 0.73, 0.67 ± 0.23, 1.93 ± 0.42, 2.12 ± 0.85, and 0.32 ± 0.06 mg/l, respectively, whereas the lower level were 0.03 ± 0.04, 0.15 ± 0.25, 0.11 ± 0.14, 0.02 ± 0.03, and 0.17 ± 0.29 mg/l, respectively, at downstream locations (Khaadi and Jajal). Therefore, the result of heavy metals revealed that the concentration was accelerated from upstream locations to downstream locations. Similar observations for the Ganga river system were recorded and found unpolluted water quality (United States Environmental Protection Agency (USEPA) 2000; Ismail 2011) (Table 2).

For the evaluation of seasonal or site-wise pollution level, comprehensive pollution index (CPI) was used and given in Table 2. The seasonal variation in CPI values can easily be illustrated by a graphical presentation given in Fig. 2. The CPI values revealed that the water quality of River Henwal at all four sites is in very poor quality as it varied from 2.11 to 4.66. According to the rating sale of CPI, these observed values were reported above 2 which is an indication that it is highly polluted. However, the less pollution load was noticed at the control site in comparison to other sites. In graphical representation, it is clearly illustrated that there was a reduction

Table 2 Water pollution at each sampling location

Sampling sites	CPI (Autumn)	CPI (Winters)	CPI (Summer)	CPI (Monsoon)	Polluted
Control	2.19	2.11	2.62	2.36	Severely
Nagni	3.98	3.16	3.38	2.30	
Khadi	3.74	4.09	4.66	4.37	
Jajal	4.11	4.55	4.34	4.58	

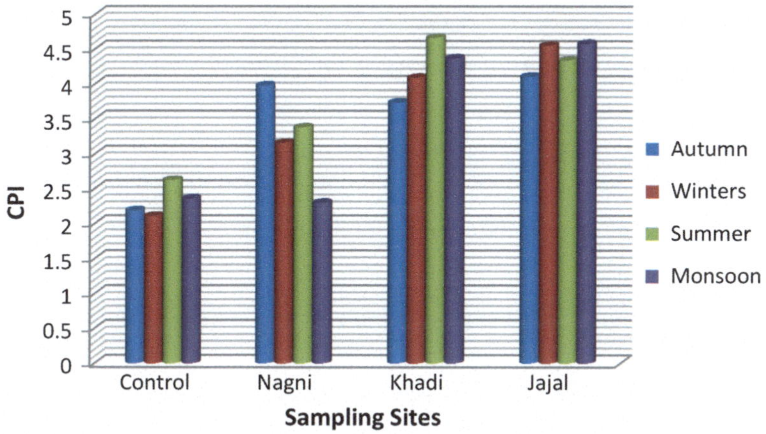

Fig. 2 Variation of CPI during autumn, winter, summer, and monsoon seasons

in the pollution level during monsoon season in comparison to the previous season, i.e., summer—it could be the reason of dilution of contaminants.

Other than the significance of phytoplanktons as the essential makers in sustenance networks and guaranteeing environmental adjustments, types of phytoplanktons can be helpful markers of water quality (Dubey et al. 2012). As illustrated in other geological zones and an assortment of natural surroundings sorts, the freshwater impact is known to have a significant impact on phytoplankton biomass, profitability, and group creation (Matta et al. 2018b). The seasonal abundance of plankton varying between upstream locations, i.e., control site to downstream locations, i.e., Jajal are represented in Table 3. Maximum plankton composition (1315.74 ± 9.04 individuals/L) was reported at Nagni during the autumn season and minimum plankton (793.63 ± 156.42 individuals/L) was reported at Jajal during monsoon. For that reason, the maximum phytoplankton density (1024.89 ± 3.69 individuals/L) at Nagni during autumn and the minimum density (508.89 ± 54.21 individuals/L) were recorded at Jajal during the monsoon season (Fig. 3a). Among phytoplanktons, different groups were identified as diatoms, green algae and blue-green algae, out of which blue-green algae were dominant and diatoms were found least throughout the study period at all four sites. The maximum density of blue-green algae was 457.75 ± 1.05 individ-

Table 3 Seasonal variation in the planktonic assemblage of four different sites

Planktons	Sites with season							
	Control site				Nagni			
	Autumn	Winter	Summer	Monsoon	Autumn	Winter	Summer	Monsoon
Total Phytoplankton	754.68 ± 2.41	915.04 ± 2.54	580.89 ± 4.33	561.33 ± 4.80	1024.89 ± 3.69	953.89 ± 2.69	895.45 ± 3.69	710.65 ± 5.16
Diatoms	261.73 ± 0.36	235.09 ± 0.53	187.42 ± 1.21	119.82 ± 2.51	251.42 ± 1.06	290.32 ± 0.21	269.02 ± 1.13	198.35 ± 2.45
Green algae	289.49 ± 0.43	298.67 ± 0.83	178.92 ± 1.56	186.84 ± 1.28	315.72 ± 1.58	292.15 ± 1.16	307.78 ± 1.25	204.76 ± 1.20
Blue Green algae	203.46 ± 1.62	381.28 ± 1.18	214.55 ± 1.56	254.67 ± 1.01	457.75 ± 1.05	371.42 ± 1.32	318.65 ± 1.31	307.54 ± 1.51
Total Zooplankton	342.87 ± 6.34	323.94 ± 4.45	287.01 ± 5.45	249.72 ± 6.84	290.85 ± 5.35	310.38 ± 6.08	281.50 ± 4.76	288.90 ± 6.36
Protozoa	95.10 ± 0.49	103.32 ± 0.51	88.64 ± 16.32	77.11 ± 1.49	86.52 ± 1.16	97.11 ± 1.25	72.37 ± 1.04	71.05 ± 1.81
Rotifera	73.65 ± 0.30	78.56 ± 1.68	60.64 ± 9.11	58.83 ± 1.56	63.25 ± 1.37	57.26 ± 1.03	69.24 ± 0.32	68.29 ± 0.87
Cladocera	68.87 ± 0.90	69.30 ± 1.25	65.26 ± 10.25	61.46 ± 1.91	72.15 ± 0.32	78.09 ± 1.02	69.81 ± 1.07	70.24 ± 1.44
Copepod	46.81 ± 1.36	50.12 ± 1.07	41.53 ± 8.43	56.90 ± 1.26	37.70 ± 1.33	39.72 ± 1.70	36.76 ± 1.17	43.26 ± 0.88
Ostracods	39.51 ± 1.40	41.57 ± 1.83	33.48 ± 6.13	32.71 ± 1.33	31.23 ± 1.17	38.20 ± 1.08	33.32 ± 1.16	36.06 ± 1.36
Total Phytoplankton	879.30 ± 2.41	948.09 ± 2.54	689.23 ± 4.80	681.60 ± 4.33	846.29 ± 132.32	813.43 ± 121.22	789.13 ± 222.19	508.89 ± 54.21

(continued)

Table 3 (continued)

Planktons	Sites with season							
	Control site				Nagni			
	Autumn	Winter	Summer	Monsoon	Autumn	Winter	Summer	Monsoon
Diatoms	298.75 ± 0.36	277.42 ± 0.53	162.91 ± 2.51	201.42 ± 1.21	228.14 ± 44.32	234.32 ± 39.89	203.13 ± 66.08	156.73 ± 21.59
Green algae	281.91 ± 0.43	301.37 ± 0.83	190.92 ± 1.28	187.35 ± 1.56	232.70 ± 54.69	247.41 ± 48.84	264.49 ± 78.85	132.74 ± 23.45
Blue Green algae	298.64 ± 1.62	369.30 ± 1.18	335.40 ± 1.01	292.83 ± 1.56	321.29 ± 54.68	331.70 ± 189.52	268.57 ± 98.37	219.42 ± 45.94
Total Zooplankton	328.88 ± 6.72	315.94 ± 5.34	234.49 ± 7.37	257.12 ± 6.94	289.44 ± 26.58	278.76 ± 23.41	289.56 ± 28.78	284.74 ± 25.69
Protozoa	113.34 ± 1.11	91.10 ± 0.45	83.27 ± 0.58	79.11 ± 1.47	88.14 ± 13.79	97.12 ± 17.82	89.34 ± 13.90	74.43 ± 12.93
Rotifera	71.56 ± 1.67	76.65 ± 1.30	49.82 ± 1.46	57.83 ± 1.96	62.72 ± 7.90	55.26 ± 6.76	51.21 ± 4.84	63.25 ± 8.02
Cladocera	66.30 ± 1.24	62.87 ± 0.81	56.04 ± 1.73	67.46 ± 1.11	66.43 ± 7.13	62.45 ± 7.01	52.34 ± 6.89	61.89 ± 7.03
Copepod	54.12 ± 1.17	47.81 ± 1.37	46.03 ± 1.24	52.90 ± 1.21	44.62 ± 6.51	41.57 ± 4.32	52.63 ± 6.31	47.91 ± 5.79
Ostracods	23.56 ± 1.53	37.51 ± 1.41	21.96 ± 1.93	35.02 ± 1.62	32.77 ± 6.06	25.65 ± 2.51	22.87 ± 2.13	41.58 ± 4.74

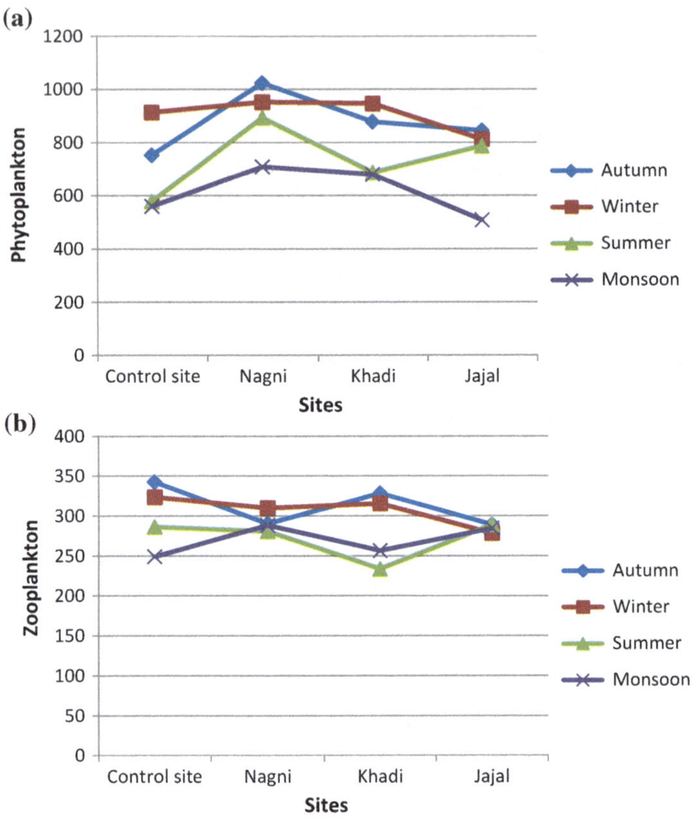

Fig. 3 Seasonal variation in phytoplankton composition (**a**), in Zooplankton composition (**b**)

ual/L at Nagni during the autumn season and the minimum abundance was 219.42 ± 45.94 individual/L at Jajal during monsoon. The depth of water level, temperature, and nutrients substances affected the growth of the phytoplanktons. The highest level of pH and bicarbonate increased the growth of blue-green algae and diatoms (Matta et al. 2018c).

Similarly, the higher density of zooplankton (342.87 ± 6.34 individuals/L) was noticed at the control site during autumn and the lower density (234.49 ± 7.37 individuals/L) at Khaadi during the summer (Fig. 3b). Five major groups of zooplanktons were identified during the study period as protozoa, rotifera, cladocera, copepoda, and ostracods. Among them, protozoa were dominant and ostracods were found least as compared to the other groups during the study period.

The Simpson's index (D) and Shannon–Wiener index (H) were calculated to observe the density and diversity of individual zooplankton groups. Group Evenness (E) was also observed to find out the distribution of individuals zooplanktonic group. The maximum value of Simpson's index, Shannon–Wiener index, and Group Evenness was 4.80, 1.59, and 0.46, respectively, at Jajal during monsoon season.

Table 4 Biotic indices for the density, diversity, and evenness of zooplankton

Biotic indices	Sites							
	Control site				Nagni			
	Autumn	Winter	Summer	Monsoon	Autumn	Winter	Summer	Monsoon
D	4.57	4.54	4.5	4.71	4.43	4.41	4.56	4.68
H	1.56	1.56	1.55	1.58	1.54	1.54	1.56	1.57
E	0.44	0.44	0.44	0.46	0.43	0.43	0.44	0.45
Biotic indices	Sites							
	Khadi				Jajal			
	Autumn	Winter	Summer	Monsoon	Autumn	Winter	Summer	Monsoon
D	4.18	4.57	4.36	4.7	4.53	4.24	4.33	4.8
H	1.51	1.56	1.53	1.58	1.56	1.52	1.53	1.59
E	0.41	0.44	0.43	0.46	0.44	0.42	0.43	0.46

The minimum value of D, H, and E was 4.18, 1.51, and 0.41, respectively, at Khaadi during the autumn season (Table 4). According to the rating scale of these biotic indices, the value of H below 3 indicates that the water quality is contaminated and above 3 indicates clean water quality.

4 Conclusion

The present study represents the pollution level of River Henwal by measuring the different hydrological parameters along with the assemblage of planktons and then calculating the CPI. The CPI of the Henwal river (2.11–4.66) indicates that the water quality is highly polluted. Just like the water quality of River Ganga is getting deteriorated by different tourist activities, dispenses of agricultural and industrial wastes, its tributary (i.e., the Henwal river) also shows the same conditions. Planktonic assemblage observed in river water indicates the presence of suitable organic and inorganic contamination. This study provides information to the government, policy-makers, and environmental conservators regarding the protection and purify of the water quality of River Henwal.

References

American Public Health Association (APHA) (2012) Standard methods of water and wastewater analysis 22th Edition. American Public Health Association. Washington DC

Badrakh A, Chultemdorji T, Tserendorj T, Vanya D, Dalaijamts C, Hagan R, Govind S, Shinee E (2008) A study of the quality and hygienic conditions of spring water in Mongolia. J Water Health 6:141–148

Bureau of Indian Standards (BIS) (2012) Indian standard specification for drinking water 10500
Census (2011). Chamba population. Retrieved from Census 2011. http://www.census2011.co.in/data/town/800301-chamba-uttarakhand.html
Chauhan A, Singh S (2010) Evaluation of Ganga water for drinking purpose by water quality index at Rishikesh, Uttarakhand India. Rep Opin 2(9):53–61
Dubey VK, Sarkar UK, Kumar RS, Mir JI, Pandey A, Lakra WS (2012) Length–weight relationships (LWRs) of 12 Indian freshwater fish species from an un-impacted tropical river of Central India (River Ken). J Appl Ichthyol 28:854–856
Environmental Protection Agency (EPA) (2012) A report on water: monitoring & assessment
Haritash AK, Gaur S, Garg S (2016) Assessment of water quality and suitability analysis of River Ganga in Rishikesh, India. Appl Water Sci 6:383–392. https://doi.org/10.1007/s13201-014-0235-1
Ismail Z (2011) Monitoring trends of nitrate, chloride and phosphate levels in an urban river. Int J Water Resour Environ Eng 3(7):132–138
Joshi DM, Bhandari NS, Kumar A, Agrawal N (2009) Statistical analysis of physicochemical parameters of water of River Ganga in Haridwar district. Rasayan J Chem 2:579–587
Kamboj N, Matta G, Bharti M, Kumar A, Kamboj V, Gautam, RK (2017) Water quality categorization using WQI in rural areas of Haridwar, India. ESSENCE Int J Env Rehab Conser 8(2):108-116
Matta G (2010) Freshwater: resources and pollution. Environ Conserv J 11(3):161–169
Matta G (2014) A study on physico-chemical Characteristics to assess the pollution status of river Ganga in Uttarakhand. J Chem Pharm Sci 7(3):210–217
Matta G (2015) Effect of water quality on phytoplankton ecology of Upper Ganga Canal. Int J Sci Eng Res 6(2):762–768
Matta G, Gjyli L (2016) Mercury, lead and arsenic: impact on environment and human health India. J Chem Pharm Sci 9(2):718–725
Matta G, Kumar A (2015) Monitoring and evaluation of river Ganga system in Himalayan region with reference to Limnological aspects. World Appl Sci J 33(2):203–212
Matta G, Kumar A (2017a) Health risk, water hygiene, science and communication. ESSENCE Int J Environ Rehabil Conserv 8(1):179–186
Matta G, Kumar A (2017b) Role of science and communication in health and hygiene: a case study. ESSENCE Int J Environ Rehabil Conserv 8(2):95–101
Matta G, Uniyal DP (2017) Assessment of species diversity and impact of pollution on limnological conditions of River Ganga. Int J Water 11(2):87–102
Matta G, Pandey RR, Saini KK (2015a) Assessment of pollution on water quality and phytoplankton diversity in canal system of River Ganga. World J Pharm Res 4(11):889–908
Matta G, Srivastava S, Pandey RR (2015b) Assessment of physicochemical characteristics of Ganga Canal water quality in Uttarakhand. Environ Dev Sustain19:419–431. https://doi.org/10.1007/s10668-015-9735-x
Matta G, Kumar A, Uniyal DP, Singh P, Kumar A, Dhingra GK, Kumar A, Naik PK, Shrivastva NG (2017) Temporal assessment using WQI of river Henwal, a tributary of river ganga in himalayan region. ESSENCE Int J Environ Rehabil Conserv VIII(1), 187–204
Matta G, Gjyli L, Kumar A, Machell J (2018a) Hydrochemical characteristics and planktonic composition assessment of River Henwal in Himalayan Region of Uttarakhand using CPI, Simpson's and Shannon-Weaver Index. J Chem Pharm Sci 11(1):122–130
Matta G, Kumar A, Kumar A, Naik PK, Kumar A, Srivastav N (2018b) Assessment of heavy metals toxicity and ecological impact on surface water quality using HPI in Ganga river. INAE Lett. https://doi.org/10.1007/s41403-018-0041-4
Matta G, Kumar A, Tiwari AK, Naik PK, Berndtsson R (2018c) HPI appraisal of concentrations of heavy metals in dynamic and static flow of Ganga River System. Environ Dev Sustain https://doi.org/10.1007/s10668-018-0182-3
Matta G, Kumar A, Dhingra GK, Singh P, Gjyli L, Kumar A (2018d) Limnological assessment of anthropogenic activities of Canal Henwal. J Chem Pharm Sci 11(1):77–86

Matta G, Naik PK, Machell J, Kumar A, Gjyli L, Tiwari AK, Kumar A (2018e) Comparative study on seasonal variation in hydro-chemical parameters of Ganga River water using comprehensive pollution index (CPI) at Rishikesh, (Uttarakhand) India. Desalination Water Treat https://doi.org/10.5004/dwt.2018.22487

Mishra S, Kumar A, Shukla P (2015) Study of water quality in Hindon River using pollution index and environmetrics, India. Desalination Water Treat https://doi.org/10.1080/19443994.2015.1098570

Rani N, Sinha RK, Prasad K, Kedia DK (2011) Assessment of temporal variation in water quality of some important rivers in middle Gangetic plains, India. Environ Monit Assess 174:401–415

Ravindra K, Meenakshi A, Rani M, Kaushik A (2003) Seasonal variations in physicochemical characteristics of river Yamuna in Haryana and its ecological best designated use. J Environ Monit 5:419–426

Shannon CE, Weaver W (1949) The mathematical theory of communication. University of Illinois Pres, Urbana, p 144

Sharma RK, Rathore V (2000) Pollution ecology with reference to commercially important fisheries prospects in rural-based water body: the Lake Sarsai Nawar, Etawah (Uttar Pradesh). Pollut Res 19:641–644

Sharma RC, Singh N, Chauhan A (2016) The influence of physico-chemical parameters on phytoplankton distribution in a head water stream of Garhwal Himalayas: a case study. Egypt J Auatic Res 42:11–21

Simpson EH (1949) Measurement of diversity. Nature 163:688

Singh P (2014) Studies on seasonal variations in physico-chemical parameters of the River Gomti (U.P.) India. Int J Adv Res 2(2):82–86

United States Environmental Protection Agency (USEPA) (2000) Newport bay toxics TMDL freshwater flow and seasonal variation. Technical support document, Part B, 1–6

Uttarakhand Tourism Development Board (2018, May 4). Chamba. Retrieved from the Uttarakhand Tourism Development Board. http://uttarakhandtourism.gov.in/chamba

World Health Organization (WHO) (2011) Guidelines for drinking water quality, 4th ed.

Zhao Y, Xia XH, Yang ZF, Wang F (2012) Assessment of water quality in Baiyangdian Lake using multivariate statistical techniques. Proc Environ Sci 13:1213–1226

The manufacturer's authorised representative in the EU is Springer Nature Customer Service Centre GmbH, Europaplatz 3, 69115 Heidelberg, Germany. If you have any concerns regarding our products, please contact ProductSafety@springernature.com

Printed and bound by CPI Group (UK) Ltd, Croydon, CR0 4YY
30/09/2026
02225458-0003